Insect Biodiversity:
Functional dynamics and Ecological Perspectives

Prof. T.N. ANANTHAKRISHNAN

(Former Director, Zoological Survey of India, Calcutta, W.B.
& Ex-Founder Director, Entomology Research Institute, Chennai, T.N.)

SCIENTIFIC
PUBLISHERS

Published by

SCIENTIFIC PUBLISHERS (INDIA)

Jodhpur –

5 A, New Pali Road
P.O. Box 91
Jodhpur - 342 001 INDIA

Delhi –

4806/24, Ansari Road
Daryaganj
New Delhi - 110 002 INDIA

Print: 2018

ISBN: 978-81-7233-641-7

Visit the Scientific Publishers (India) website at
http://www.scientificpub.com
Printed in India

"A greening of the human mind must precede the greening of the earth. A green mind is one that cares, saves and shares. These are qualities essential for conserving Biodiversity, now and for ever."

— M.S. Swaminathan, FRS

Preface

A knowledge of the ecological basis of biodiversity is an important element of paramount importance for the preservation of genetic diversity and sustainable utilisation. Preserving habitats as a means to protect threatened species, with such activities tending to lead to the acceleration of biodiversity becomes a useful criterion. All species deserve adequate recognition regardless of their utilitarian value, in view of their being components of our life support system. Species richness or the number of species at a site or habitat, be it an agroecosystem or a forest habitat is an essential criterion. It is in this connection that the present edited volume seeks to identify species diversity in a host of ecosystems with a view to assess the wealth of species, notably in agroecosystems, natural forest ecosystem, forest floor, beverage and spice crops, desert ecosystem and of vector insects. Needless to emphasize that the ecological resources and services are generated by the consistent interaction between organisms, population, community and their physical and chemical environment. Species diversity increases initially and then becomes stabilized or declines in later stages of ecosystem development. In recent years emphasis is laid on role of ecological scenario in the meaningful management of bioresources. An important area wherein biodiversity studies have been consistent, is the concern over the loss of forest ecosystems and resultant implication on biodiversity. The ecological and economic implication of these changes appear very relevant, notably forest loss and resultant effect on species composition and extinction. The chapters on social forestry and natural forestry add toour knowledge of biodiversity, besides the beverage and spice crops. The impact of temperature on biodiversity in such ecosystems as the hot desert and cold desert ecosystems further add to our existing knowledge of the ecological dimensions of biodiversity. Last but not least, the impact of closed ecosystems as occurring in plants galls provides an added dimension of ecosystems and the impact of gall diversity in insect biodiversity.

The diversity involved in environmental degradation with increasing conversions and alterations of habitats is of commonplace occurrence, although the establishment of plantations and secondary forests have compensated for the overall loss of biodiversity. Stability in the structure and composition involving populations, communities and landscape needs to be assured. The decline of populations and directional change in composition is often used as a criterion in their conservation. ``Loss of species and continued depletion of population with habitat simplification may eventually lead to the breakdown of ecological function, thereby causing ecosystem collapse". Therefore, the major challenge to biodiversity conversion is to sustain ecosystem functioning and maintain resilience of ecosystem to change".

The need for top quality taxomonic revision has never been greater than it is today, involving new ways to apply systematic knowledge to conservation problems. One cannot overlook the need for the application of systematics in the conservation of biodiversity. Survival of biodiversity is a prerequiste for the survival of the ecological and loss of living components in an ecosystem will leadto changes in the local environment. Needless to emphasize that development programmes should be planned and executed, taking into consideration their impact on biodiversityand toachieve their continued monitoring conservation of biodiversity is vital.

My sincere thanks are due to everyone of the authors of the different chapters whose expertise over the years has contributed to the realisation of the need for a better understanding of biodiversity and become aware of the complexities involved in the study of biodiversity.

Chennai
1-8-2009

T.N. Ananthakrishnan

Contents

Chapter - 1

Perspectives of Changing Biodiversity : Ecological and Economic Implications

T.N. Ananthakrishnan

Flat 6, Dwaraka, 22, Kamdarnagar Ist, Chennai- 600 034

Introduction

Increased realization of the deterioration of biological systems over the last couple of decades has led to a better appreciation of the loss of genes, species and ecosystems. Biodiversity has assumed increasing relevance today and ever since 1992, the year of the International Convention of Biodiversity in Rio de Janeiro, it has come to occupy the central stage, holding the key for the maintenance of life. It is a unifying discipline bringing together the ecologist, economist, entomologist, educationalist and the environmentalist, resulting in an interdisciplinary, multifunctional subject, involving ecosystem dynamics, environmental and climatic change, energy resources, biodiversity and sustainable development at various levels-local, national and global. In short, biodiversity involves the totality of genes, species and ecosystems which influence productivity and services rendered, such as soil fertility, maintenance, water management and pest management. Biodiversity operates at various levels in the ecosystem from the lowest sub specific variation in natural populations to ecological entities. As presently realized biodiversity refers to species richness and abundance, genetic variability between and within populations, ecosystems and community diversity. To the ecologist in particular, biodiversity may include the number of trophic levels, guilds and diversity of life cycles. The need for a greater appreciation of biodiversity has arisen today in view of increasing habitat loss, fragmentation and modification we see all around us, which has led to the disappearance of several species and in terms of biodiversity, both the loss of species and gene reservoirs pose a serious threat to human welfare. Habitat fragmentation results in the sub-division of habitats into smaller units resulting in their increased insularity as well as loss of total habitat area. Such fragmentation changes the microenvironment at the fragmented edges and edge effects include microclimate, changes in light, temperature, wind and humidity and each of these can have a significant impact on the vitality and composition of the species in the fragment. Habitat fragmentation is a leading cause for biodiversity loss and has further increased the local rate of extinction, by reducing population sizes and by changing the physical and biological environments through edge effects. Community or ecosystem processes are affected to such a degree, that they have disproportionately large influence on the character of ecosystems. Needless to emphasize that phenotypic flexibility, genetic variation within population, ecotypic variation, functional diversity, community diversity, gradient diversity, ecological diversity, besides species richness and diversity are essential ingredients of biodiversity.

Changing landscapes, ecodiversity and insect faunal dynamics

Substantial changes in vegetation, notable in some areas of the Western Ghats like the Tenmalai region of Kerala, over a considerably long period have resulted in extensive alteration in the landscape profiles, leading to changes in insect biodiversity with the loss of specific mix of species from sites where they were once abundant. Heterogeneity of an area is strongly correlated with the number of species in the area and patterns of species diversity are associated with patterns of spatial and temporal variation. Stable ecosystems, diversity of habitats, abundant biomass and diversity of plant and animal species, undoubtedly add to species diversity and increased denudation of natural forests and replacement by monoculture, have resulted in reduced insect abundance and species diversity. An understanding of the spatial distribution of biotic elements is an important aspect to assess changes in the ecological structure, composition and function in response to natural as well as disturbed conditions. While biodiversity essentially deals with diversity of genes in populations in an ecosystem and diversity of ecosystems in a landscape, preservation of biodiversity demands information on the units of biodiversity in relation to different microsites, not to mention of the nature of variation in diversity. This calls for an intensive niche-oriented exercise with assemblages in relation to the coexistence of different species at the survey sites. With virtually no concentrated effort being made to involve entomological studies relating to the dynamics of the canopy, bark, litter, galls, rhizosphere, epiphytes, inflorescence, pteridophytes and natural enemy complexes involved in forests such as those of the Western and Eastern ghats, the gap in our knowledge is indeed wide. It is here that expertise is called for on diverse group of insects to be able to identify the microhabitats harbouring the insects, instead of random efforts to collect them in isolation. The exuberance of such microhabitats in forest ecosystems is awe-inspiring as is the complexity with which natural communities are assembled and there can be no better examples than the diversity of natural forest litter community which harbours complex food webs, and more diverse the plant community, more diverse will be the litter community. My own limited experience working in some areas of the Western Ghats over the years has enabled plant galls which are often abundant and readily available system for ecological studies, so clearly delineated in the chapter on galls by Raman. Communities of insects and mites associated with galls of certain plants tend to be of considerable ecological interest. These inter relationships are intricate and challenging. Galls of each species are so constant in structure, usually differing conspicuously from galls of related species growing on the same plant. It is therefore easy to conclude that a key specific, precise mechanism is employed by each gall former to manipulate plant growth, with both hypertrophy and hyperplasy contributing to the formation of most galls. Identification of diverse microsites with Indo-Malayan, Ethiopian and Neotropical elements, some of which are good examples of discontinuous distribution, besides sex-limited polymorphism showing unbelievable phenotypic flexibility resulting in endless patterns deserves scrutiny.

Needless to emphasize that the patterns of insect diversity reflect the structural role of plants in creating environmental heterogeneity. Such heterogeneity has enabled the persistence of high diversity among the diverse species of insects which directly or indirectly depends on plants. The diversity of caterpillars associated with plants is an important aspect to be able to confidently assess the nature of butterfly diversity. The development of environmental versatility is an important criterion in the successful exploration of various habitats by insects and in the process insects develops structural, physiological and behavioural diversities enabling them to respond differently to diverse habitats at different periods of development and under varying conditions. Plants offer microenvironments necessary for a variety of functions. The fluctuations in the habitat combined with the internal environmental factors have positively stimulated the evolution of structurally diverse series of forms. Diversity of resources available in a community determines the variety of opportunities for ecological diversification. Of equal significance is the fact that genetic variability among individuals of the same species can impose differences in size, morphology, behaviour and other phenotypic traits independent of environment conditions besides limiting the range of phenotypic flexibility that can occur. The unbelievably large diversity of ants, membracids, thrips, earwigs, beetles to mention a few, as well as trophic relationships at the various microsites are aspects worth immediate attention.

Ground surface ecosystem and litter dynamics

Beyond the species level, diversity includes the patterns in nature upto the landscape level. The occurrence of continuously changing mosaics from virgin or undisturbed forest patches and denuded patches to these rcovering from the disturbance is a typical present day phenomenon, so that spatial and temporal patterns of differences become evident in the landscape. Denudation of natural forests and replacement by monoculture such as teak, rubber etc., results in mortality from different biotic and abiotic causes, reducing insect abundance especially in the forest floor, litter dynamics making a useful study. Periodic burning in several places in the Western Ghats involving the removal of accumulated litter results in the substantial loss of organic matter affecting the faunal communities. Diversity of species is not only determined by the type of litter, but also as a function of the complex abiotic and biotic components related to the heterogeneity of the litter. Further, the chemical nature of the litter involving the inorganic components as well as the phenolics, volatiles and flavonoids also contribute considerably to the degree of complexity of the litter insect community (Ananthakrishnan, 1996). The species composition in natural forest litter also reveals the existence of potential complicated trophic links. The impact of human interference has therefore considerable bearing on litter insects guilds, resulting in their fragmentation and ultimate disappearance. Species richness is high in natural litter of tropical communities, the micro-arthropod populations varying in direct relation to the amount of litter even in monoculture forest like *Shorea robusta*, *Tectona grandis* and *Eucalyptus*. As the litter layer becomes deeper the successional seres become very evident and as the litter undergoes an orderly and sequential breakdown, the microfloral and faunal assemblages tend to succeed one another. Biodiversity of arthropods can be related to the morphology of litter breakdown with Protura. Diptura and Collembola being the major soil dwellers, Collembola being the most predominant. The type of soil composition of microflora, soil moisture and the type of vegetation influence the abundance of Collembola. Besides several species of mycophagous thrips inhabit fungus infested surface litter in natural forests wherever there is rich accumulation of litter. Many mycophagous species are associated with particular types of fungal infestation in litter and instances are known where one species may feed on the specific spores of a fungus or many polyphagous species feed on a wide variety of fungi. Most of the populations of these litter inhabiting mycophagous thrips species show phenotypic plasticity resulting in a variety of phenotypes, such variability enabling exploitation of food resources resulting in the increase of breeding potential (Ananthakrishnan, 1996).

Changing mosaics and patch dynamics

While the protection of terrestrial vegetation is positively correlated with plant species diversity, variation in physiological and biological patterns influences diversity. Heterogeneity is contributed by plants in diverse ways involving vertical structure and complexity produced by the roots, stems and leaves and ground vegetation. Beyond the species level, diversity includes patterns in nature up to the landscape level. The occurrence of continually changing mosaic from virgin or undisturbed forest patches and denuded patches to patches recovering from disturbance is a typical present day phenomenon, so that spatial and temporal patterns of differences are evident in the landscape. Maximum diversity within a local patch will be influence by the number of species available in the landscape, i.e. the regional species pool for dispersal into the patch. Denudation of natural forests and replacement by monoculture results in mortality from different biotic and abiotic causes, reducing insect abundance especially in the forest floor. Large scale clearing results in the break-up of recycling mechanisms leading to a rapid loss of accumulated nutrient capital. Once it is lost, vegetation may not recover its former diversity or stature and regeneration will not involve the same species diversity. Local extinction of the species is thus an inevitable consequence of such disturbances.

Heterogeneity of an area is strongly correlated with the number of species in the area and patterns of species diversity are associated with patterns of spatial and temporal variation. Stable ecosystem, diversity of habitats, abundant biomass and diversity of plant and animal species undoubtedly add to species diversity. Increased denudation of natural forests and replacement by monoculture result in the reduced abundance and species diversity. Another dimension of biodiversity relates to agro biodiversity, wherein human manipulation and alteration of ecosystems for the purpose of increasing agricultural production, make agroecosystem structure and functioning very different from natural ecosystems. The spread of modern agriculture has resulted in tremendous changes in landscape diversity with increase in crop diversity. The spread of high yielding crop varieties has had devastating effects on the genetic diversity of food plants. Several local rice varieties have becomes extinct over the past few years with biodiversity tending to face greater long term risks.

High rate of population growth in countries such as Bangladesh, have resulted in increasing demand for food, by increasing cropping intensity and irrigated land, introduction of modern crop varieties and increasing fields. All this can be achieved only with further degradation of agricultural resource base. Land and soil degradation are inevitable due to overuse and misuse of agricultural resources, with soil organic matter becoming reduced by acute biomass crisis and use of crop residues. Increased use, diversion and loss of net primary productivity, result in high rates of loss of biodiversity.

Ecosystem sustainability : role of biodiversity

Human welfare depends on climate stability, watershed protection and erosion control. Biodiversity addresses the interaction between and economic systems to determine the exact level of exploitation. Our societies largely depend on ecological sources for welfare and resilience which contributes to ecosystem integrity. The need for a better appreciation of the interaction between ecological and economic systems to determine the exact level of exploitation, cannot be over emphasized. Resilience contributes to ecosystem integrity and the need for a better appreciation of this with in biological communities appears relevant, so that investigations of critical thresholds of diversity become important. For a better appreciation of ecosystem function, the role of individuals, species, their number and functional competence in their natural habitat and changes in the dynamics of natural populations appear relevant. A critical mix of species relating to the maintenance of eocystems becomes relevant for an assessment of the functional dynamics. What is significant is a better appreciation of the self organizing ability of ecosystems. Generation of ecological services and maintenance of resilience of ecosystems to changes, play a useful role in the maintenance of ecosystems. Ecosystem services refer to the diverse condition and processes associated with natural ecosystem that confer some benefits to humanity. To aid their delivery the need for the right mix of biodiversity cannot be over emphasized. Some of the more important instances of ecosystem services relate to generation and maintenance of fertile soil; the regulation of the hydrological cycle; amelioration of climate, cycling of nutrient and water, detoxification of waste products and preservation of earth's genetic library viz. its agricultural, medicinal and important element of industrial enterprise (Ehrlich, 1977). For persistence of life on earth, the role of microbial activity in completing the nutrient cycle in individual ecosystems and biosphere as a whole is critical. The number, functions and activities of individual species in their microhabitats and the role of species composition and their dynamics are necessary inputs in preserving the functional integrity of the system. As such the primary ecological importance of biodiversity must be its role in preserving ecosystem resilience.

Ecological implications of biodiversity

The ecological implications of biodiversity become important since biodiversity at the genetic, specific population and ecosystem levels contribute in maintaining these resources, Each ecosystem is composed of a number of physical, biological and chemical components such as soil, water, plant and animal species and nutrients : Ecological functions such as biological productivity, nutrient cycling, hydrology are of importance in maintaining the overall performance and integrity of the ecosystem, enabling supporting and sustaining the various living organism on it. These ecological are often referred to as ecological services and the components of the biological services ecological resources. The need for a knowledge of the role of particular species cannot over emphasized and human societies are dependent on ecological services which are generated by the proper functioning of ecosystems. Biodiversity in short, plays an important role in ecological functioning. There is a minimal diversity for ecosystem functioning and adding or deleting species has no noticeable effect. More species rich assemblages have greater productivity due to their ability to exploit a greater variety of resources. Intercropping in agriculture and agroforestry exploit this phenomenon (Lawton, 2000). A species with diverse population is more likely to include individuals that are suited to new conditions than a single local population (Barbier, Burger and Folke, 1995).

In short, phenotypic flexibility, genetic variation within populations, ecotypic variation, functional diversity, community diversity, gradient diversity, landscape diversity, besides species richness and diversity are essential ingredients of biodiversity. Some species play a more significant role in the ecosystem than other as indicators of ecological processes influencing community structure. Needless to emphasize that ecological and economic monitoring of threatened species and ecosystems need to be realized notably the role of biodiversity in ecological functioning and resilience. Basic information on the consequences of biodiversity loss must be provided to enable decision making in conservation and development (Barbier *et al.*,1996)

Four kinds of habitat loss are recognized by Hanski (2005) *viz.* loss of quality, of quantity, of connectivity, and loss of continuity. Needless to emphasize that the greatest threat is the loss of microhabitats. Structural simplification of macrohabitats means loss of microhabitats which has become a major concern, particularly when all land and habitats are under human influence. The scale or size of human population expansion is an important aspect of the sustainability of ecosystem through reducing biodiversity and loss of resilience in the ecosystem due to the ecosystems exceeding their carrying capacity. In the context of biodiversity conservation, the need for sustaining the amount of biodiversity which would ensure the resilience of ecosystems enabling the flow of important renewable resources and ecological services to human communities becomes important. In short, conservation of biodiversity is of vital importance since some degree of biodiversity is essential for the functioning of ecosystems An effective monitoring programme utilizing a variety of indicators to assess environmental responses at population species community and ecosystem levels become a useful exercise. A species may become more abundant following habitat restoration. Declining species tend to recover leading to expected population size. Species already regionally extinct tend to return to their habitat after the quality is improved. Successful habitat restoration leads to the appearance of new habitats at a site where no suitable habitat existed before (Hanski, 2006). The sensitivity of many insect species to environmental impact including fragmentation, habitat modfication, ecological disruption, climate change provide a useful exercise to study the ecological health of microhabitats. The general trends and the end result in the most degraded habitat is a drastic reduction of species richness. While the majority of species may decline and disappear with disturbance, others increase and prosper generating a turnover of species within assemblages. Climate change will also introduce new species and delete others from regional pools, local species richness tending to change much more easily. Virtually nothing is known about how global change will alter the vulnerability of communities to invasion (Lawton, 2000). In this connection mention may be made of invasive species which pose a major threat to global change. As climatic change progresses opportunities for invasive species tend to increase. The impact of these invasive tend to be great completely altering the character of certain ecosystems often resulting in considerable damage, driving native species to extinction. The need for increased education on invasive species cannot be overemphasized, involving factors which determine whether a species will be an invasive one or not. An intensive effort is needed on the part of conservation biologists, ecologists and taxonomists in the development of new methodologies using terrestrial insects and other arthropods as indicators for an assessment of conservation and monitoring.

Keeping these aspects in mind the book seeks to provide information on biodiversity in diverse ecosystems such as biodiversity in social forests; in scrub jungles, biodiversity of insect biocontrol agents, of cecidogenores insects and in cold and hot deserts. Needless to emphasize that building an understanding of biodiversity and ecosystem functioning requires a universal body of knowledge which in turn requires continuous research exploration and publication in order to recognize that promoting increased taxonomic knowledge is a worthy goal to advance our understanding of biodiversity.

References

Ananthakrishnan, T.N. 1973. Micro arthropods and soil ecosystem. *J. Bombay Nat. Hist. Soc.*75(3): 625-61.

Ananthakrishnan, T.N. 1984. *Bioecology ot thrips.* Indira Publishing House Michigan.

Ananthakrishnan, T.N. 1986. *Bioresources exology.* Oxford and IBH, New Delhi,172 pp.

Ananthakrishnan, T.N. 1988. Dynamics of litter ecosystems in natural and interference forest. *Int. J. Ecol. Envir. Sci.,* 14 : 61-66.

Ananthakrishnan, T.N. 1992. *Dimensions in insects plant interaction.* Oxford IBH, New Delhi.

Ananthakrishnan, T.N. 1994. The endless limits of insect diversity. *Curr. Sci.* 66(4) 271-81.

Ananthakrishnan, T.N. 1996. *Forest litter communities : Biolgy and Chemical Ecology.* Science Publishers, Inc. Labanon, NH 03766,USA, 168pp.

Ananthakrishnan, T.N. 2000. Changing landscape, ecodiversity and insect faunal dynamics: the Tenmalai (Western Ghats) experience. *Curr. Sci.,* 78(4) : 387-390.

Ananthakrishnan, T.N. 2002. Insects, plants and molecular interaction. Madras Science foundation, 120pp.

Ananthakrishnan, T.N. 2004. Changing dimensions of entomological research: need for a holistic approach. *Proc. Nat. Symp. on Frontier areas of entomological Research* 2003, IARI, New Delhi,: 1-12.

Ananthakrishnan, T.N. and Douglas Whitman 2005. *Insect phenotypic Plasticity.* Science Publishers, Lebanon, USA.

Ananthakrishnan, T.N. and Sabu Thomas, K. 1993. Chemical ecology of a few heterogenous forest litter ecosystems and related trophic interactions of insect communities, *Int. J. Ecol. Envir. Sci.,* 19 : 143-160.

Ananthakrishnan, T.N. and K.G. Sivaramakrishnan 2006. *Animal Bio-diversity: Patterns and Processes,* Scientific Publishers (India), Jodhpur.

Annadurai, R.S. Chandrasekhar, S.S. and Balu, A 1988. Trophic Structure and diversity in some litter inhabiting microarthropods of monoculture and natural forest ecosystem. *Proc. Indian Acad. Sci., (Anim. Sci.)* 97 (4) : 301-308.

Barbier, E.C., Burger, J.C. and Folke, C. 1996. *Paradise lost: Earthscan publishers, London.*

Ehrlich, P.R. 1997. *A world of wound: Ecologists and the human dilemma,* Ecology institute, oldendorf, Germany.

Harringoton, R, N.E. Stork 1995. Insects in a changing Environment, Academic Press London.

Lanski, I. 2005. The Shrinking World : Ecological consequences. Ecological institute, Nordbunte, Oldendorf , Germany.

Lawton, J.H. 2000. *Community ecology in a changing world.* Ecological Institute, Nordbunte, Oldendorf, Germany.

Ramakrishnan, P.S. 1991. *Ecology and biological invasives in the tropics.* Nat. Inst. Ecology and Int. Science Publishers, New Delhi 195 pp.

Ramakrishnan, P.S. 2002. Coping with uncertainities: Role of traditional ecological knowledge in socio-ecological landscape management, *Proc. Indian Nat Sci. Acad.* (PINSA), B 68,3: 235-254.

Samways, M.J. 1994. *Insect Conservation Biology,* Chapman Hall, London.

Singh, J.S. and E. Khurana 2002. Paradigms of Biodiversity: An Overview. *Proc. Indian Nat. Sci. Acad.* (PINSA), B 68,3: 273-296.

Terborgh, J. 1992. Diversity and the tropical Rain forest. *Sci. Amer.* New York 242 pp.

Wilson, E.O. 1998. *Biodiversity,* National Academy Press, New York.

Wilson, E.O. 1992. *The Diversity of life.* W.W. Norton & company New York.

Chapter - 2

Biodiversity of Ground Surface Dwelling Arthropods in the moist forests of the Western Ghats

Sabu K. Thomas and Shiju T. Raj

PG and Research Department of Zoology, St. Joseph's College, Devagiri, Calicut, Kerala- 673 008. email: sabukthomas@gmail.com

INTRODUCTION

Forest floor represents the uppermost organic and organic enriched mineral soil horizons, formed by the deposition of senescent leaves, twigs, larger branches, bark, flowers, fruits, spores etc and its subsequent biologically mediated decomposition. Forest floor is harboured by hundreds of arthropods and is a compressed region of intense biological activity. While the forest floor may not receive the same degree of attention as other components of an ecosystem, its health is of great importance from the view point of biological diversity, food chains and ecosystem processes (Ananthakrishnan, 1996, Hattenschwiler *et al.,* 2005, Sabo *et al.,* 2005). A remarkably diverse assemblage of ground surface dwelling arthropods, live in the litter layer and in the upper strata of the soil. These organisms are an important component of the forest ecosystem, because of their relevant role as regulators of key functional processes, such as organic matter decomposition, mineralization, nutrient cycling and soil formation (Seastedt, 1984, Hattenschwiler *et al.,* 2005) and as a vital link in the food chain as decomposers between the green world of primary producers and the brown world of detritus. Additionally, ground surface dwelling arthropods contribute to the dispersal of seeds and spores, provide prey populations for ground foraging vertebrates and recycle their dung (Frith & Frith, 1990, Andresen, 2002, Vulinec, 2002).

Study of ground surface dwelling arthropods in forests is particularly bewildering due to the peculiarities of the habitat and the diversity of its dwellers. Soil and litter ecologists struggle with the immense diversity of species about which so little is known and taxonomic constraints makes species identifications difficult. Therefore, to make the categorization of forest floor arthropods easier, they are grouped based on their function in the ecosystem into fungivore, detritivore, omnivore, herbivore and predator functional groups (Ananthakrishnan, 1996, Moore & Walter, 1988, Chen & Wise, 1999). However, the limited amount of evidence of dietary requirements and biology at various life stages of most taxa and the omnivorous feeding habits of many groups makes categorization of all ground surface dwelling arthropods into well-defined functional guilds nearly impossible (Hunt *et al.,* 1987, Lee & Pankhurst, 1992, Giller, 1996, Coll & Guershon, 2002).

There is mounting evidence that vegetation type along with rainfall pattern and topography form important drivers of the abundance of ground surface dwelling arthropods in tropical forests (Boinski & Fowler, 1989, Frith & Frith, 1990, Burgess *et al.,* 1999) due to the link between vegetation type and the quality and quantity of litter inputs to the forest floor (Swamy & Proctor, 1994, Clark *et al.,* 2003, Wood *et al.,* 2005). Variations in the leaf life span of deciduous and evergreen tree species leads to differences in the physical and chemical quality of leaf litter. Litter depth peaks perhaps once or twice through the year in deciduous forests whereas there would be no significant variations of litter depth in evergreen forests (Ananthakrishnan, 2000; Santiago *et al.,* 2005; Anu & Sabu, 2007). The Western Ghats, a global hot spot of biodiversity and the only tropical forest ecoregion of the Indian peninsula is well known for regional variation in vegetation, rainfall patterns and topography (Nair 1991; Myers *et al.,* 2000). While there are a few random studies regarding biodiversity of ground surface dwelling arthropods in single forest ecosystems in the Western Ghats (Ananthakrishnan, 1996; Anu *et al.,* 2009), no comprehensive report exists on the way arthropod communities show preferences to particular forest vegetation type and their functional group classification in different forest vegetation types. Understanding the assemblage structure and relative abundance of forest floor arthropod communities in different forest vegetation types is a prerequisite for analyzing their functioning at the ecosystem level in the Western Ghats. This chapter summarizes data on the abundance and functional group composition of ground surface dwelling arthropods in the moist south Western Ghats forests based on the faunal inventories conducted at different seasons in the moist deciduous, evergreen and montane shola evergreen forests employing Berlese extraction methods (Sabu 2009, Sabu & Shiju (in press)) and functional group classification based on Moore and Walter (1988), Borror *et al.*, (1996) and Chen and Wise (1999). High proportion of the undescribed and unknown ground surface dwelling arthropod taxa and the problems associated with insufficient taxonomic knowledge in the subcontinent makes species level assessment of taxa in the Western Ghats extremely difficult. Moreover, shift from species centered conservation towards conservation and management of communities, landscapes and ecosystems as one of the principal directions of conservation biology in recent times (Hobbs, 2004) does not warrant species level consideration of all taxa especially the "black hole groups" (New, 1999) which are difficult to study and recognize.

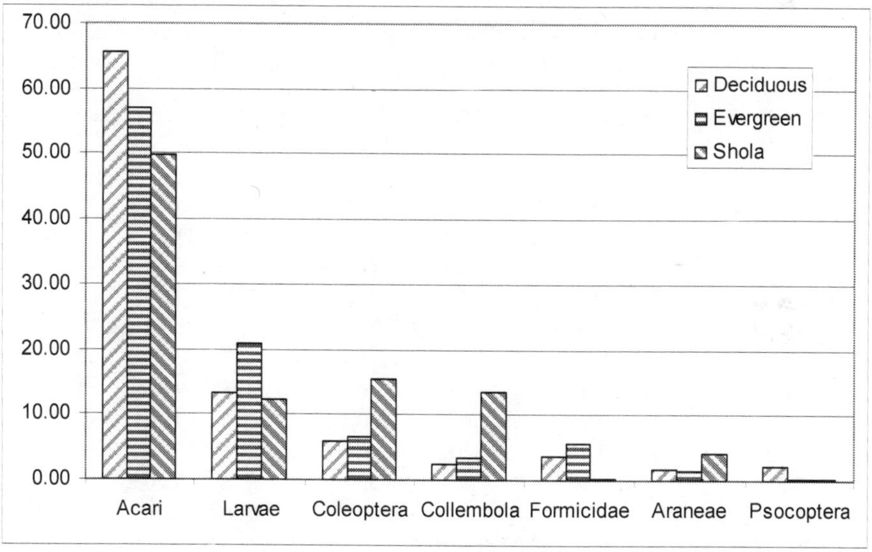

Figure 1: Proportional abundance (%) of major ground surface dwelling arthropods in deciduous, evergreen and montane shola evergreen forests in the Western Ghats.

Results

Faunal composition

Arthropods belonging to 23 orders were collected from the 3 major forest types (Table 1). Overall faunal abundance did not vary between the forest types. Six groups (Psocoptera, Hemiptera, Lepidoptera, insect larvae, Ticks (Parasitiformes) and Formicidae) showed significant variations among the different forest vegetation types. Overall faunal abundance was high in deciduous and evergreen forests and low in montane shola forest (Figure 1). Five major groups (taxa with a mean abundance of >1 and relative abundance of >2%), Acari >Insect larvae >Coleoptera >Formicidae >Collembola collectively accounted for >85 % of the total individuals captured in both evergreen and deciduous forests. Acari >Coleoptera >Collembola >insect larvae >Araneae accounted for >90% of the total fauna in montane shola forest. Proportionate distribution pattern of taxa in deciduous and evergreen forests displayed similar patterns. Distribution pattern in montane shola forests was different from other two forest types with low abundance of Formicidae and high abundance of Araneae.

Table 1. Absolute (mean ± SD) and proportional abundance (%) of ground dwelling arthropods (0.50 m^2) in deciduous, evergreen and montane shola evergreen forests in the Western Ghats.

	Arthropod groups	Deciduous forest		Evergreen forest		Shola forest	
		Mean ± SD	%	Mean ± SD	%	Mean ± SD	%
1	Acari	29.65 ± 47.94	65.70	36.45 ± 55.38	57.00	15.55 ± 26.15	49.76
2	Insect larvae	6.00 ± 6.12	13.30	13.30 ± 10.17	20.80	3.85 ± 3.67	12.32
3	Coleoptera	2.70 ± 3.59	6.00	4.25 ± 5.02	6.70	4.80 ± 7.14	15.36
4	Collembola	1.10 ± 2.15	2.40	2.20 ± 4.06	3.40	4.20 ± 7.51	13.44
5	Formicidae	1.65 ± 2.83	3.70	3.65 ± 7.86	5.70	0.05 ± 0.22	0.16
6	Araneae	0.80 ± 1.11	1.80	0.95 ± 1.23	1.50	1.30 ± 1.56	4.16
7	Psocoptera	1.05 ± 2.44	2.30	0.15 ± 0.49	0.20	0.05 ± 0.22	0.16
8	Hemiptera	0.45 ± 0.94	1.00	0.50 ± 0.61	0.80	0.05 ± 0.22	0.16
9	Diptera	0.05 ± 0.22	0.10	0.90 ± 2.40	1.40	0.00 ± 0.00	0.00
10	Thysanoptera	0.25 ± 0.79	0.60	0.20 ± 0.52	0.30	0.20 ± 0.41	0.64
11	Chalcidae	0.40 ± 1.14	0.90	0.25 ± 0.44	0.40	0.00 ± 0.00	0.00
12	Lepidoptera	0.00 ± 0.00	0.00	0.60 ± 1.60	0.90	0.00 ± 0.00	0.00
13	Ticks	0.10 ± 0.31	0.20	0.05 ± 0.22	0.10	0.45 ± 0.76	1.44
14	Chilopoda	0.40 ± 1.10	0.90	0.00 ± 0.00	0.00	0.20 ± 0.52	0.64
15	Pseudoscorpionidae	0.05 ± 0.22	0.10	0.10 ± 0.31	0.20	0.30 ± 0.73	0.96
16	Orthoptera	0.05 ± 0.22	0.10	0.10 ± .031	0.20	0.05 ± 0.22	0.16
17	Isoptera	0.20 ± 0.62	0.40	0.00 ± 0.00	0.00	0.00 ± 0.00	0.00
18	Dermaptera	0.05 ± 0.22	0.10	0.05 ± 0.22	0.10	0.05 ± 0.22	0.16
19	Blattaria	0.10 ± 0.31	0.20	0.05 ± 0.22	0.10	0.00 ± 0.00	0.00
20	Hymenoptera	0.05 ± 0.22	0.10	0.00 ± 0.00	0.00	0.05 ± 0.22	0.16
21	Diplopoda	0.05 ± 0.22	0.10	0.05 ± 0.22	0.10	0.00 ± 0.00	0.00
22	Isopoda	0.00 ± 0.00	0.00	0.10 ± 0.31	0.20	0.00 ± 0.00	0.00
23	Thysanura	0.00 ± 0.00	0.00	0.00 ± 0.00	0.00	0.05 ± 0.22	0.16
24	Homoptera	0.00 ± 0.00	0.00	0.00 ± 0.00	0.00	0.05 ± 0.22	0.16
	Total	45.2 ± 53.9		63.9 ± 60.5		31.3 ± 42.7	

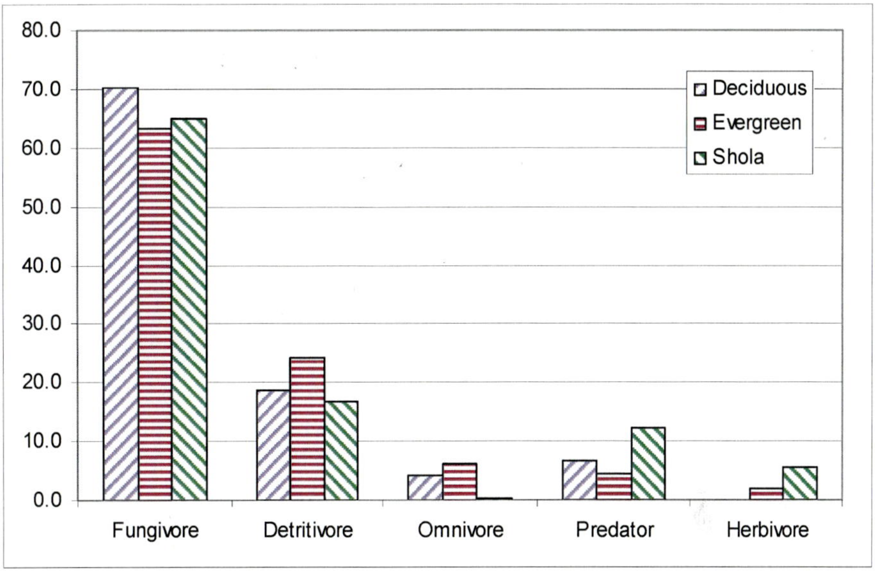

Figure 2. Functional guild composition (%) of major ground surface dwelling arthropods in deciduous, evergreen and montane shola evergreen forests in the Western Ghats

Functional Guild composition

Fungivores and detritivores functional guilds dominated all forest vegetation types. Abundance of detritivores, omnivores and herbivores varied significantly among the forests and of predators and fungivores did not (Table 1, Figures 2-3). Abundance of omnivores was lowest in Shola forests. Highest abundance of detritivores was in evergreen forests and was of same level in montane shola and deciduous forests. Abundance of herbivores was of same level in montane and evergreen forests and was absent in deciduous forests. Herbivores were the lowest represented group in evergreen and deciduous forests and omnivores in montane shola forest. High proportionate abundance of predators and low abundance of omnivores in montane shola forest was distinct. Functional guild composition was in the order of Fungivores >Detritivores >Predators >Omnivores >Herbivores in deciduous forest, Fungivores >Detritivores >Omnivores >Predators >Herbivores in evergreen forest, and Fungivores >Detritivores >Predators >Herbivores >Omnivores in montane shola forest.

Figure 3. Prominent ground dwelling beetle families in montane shola evergreen forests at Eravikulam (a. Cleridae, b. Hydraenidae, c. Hydrophilidae, d. Leiodidae), and carabid beetles (e. *Clivina lobata*, f. *Helluodes* sp., g. *Omphra pilosa*, h. *Oxylobus asperulus*, i. *Scarites ceylonicus*) in evergreen and deciduous forests in the moist south Western Ghats.

Fungivores: Acari and Collembola were the prominent fungivores in all 3 forest types. Proportionate abundance of Collembola was distinctly higher in montane shola forest than in other two forests types. Proportional distribution of the Fungivores was in the order of Acari >Collembola >Coleoptera >Diptera in deciduous forest, Acari >Collembola >Diptera > Coleoptera in evergreen forest and Acari >Collembola > Coleoptera in montane shola forests. *Detritivores:* All vegetation types showed similar pattern in the proportionate abundance of detritivores with insect larvae followed by Coleoptera as prominent groups (>95%). Isoptera, Isopoda, Diplopoda and Thysanura were the minor components of the guild. Proportional abundance of the guild was in the order of Insect larvae >Coleoptera >Isoptera >Diplopoda in deciduous forests, Insect larvae >Coleoptera >Isopoda >Diplopoda in evergreen forests and, Insect larvae >Coleoptera >Thysanura in montane shola forests. *Predators:* Coleoptera and Araneae constituted the prominent predator component of the forest floor arthropods in all vegetation types. Montane shola forests differ from deciduous and evergreen forest forests in the comparatively high proportionate abundance of Araneae and ticks. Proportional distribution of predator groups was in the order of Coleoptera >Araneae >Chilopoda >Pseudoscorpions >Ticks in deciduous forest, Coleoptera >Araneae >Pseudoscorpions >Ticks in evergreen forest and Coleoptera >Araneae >Ticks >Pseudoscorpions in montane shola forests. *Omnivores:* Lowest abundance of omnivore was in montane shola forest. Formicidae (>90%) constituted the prominent fraction of the guild in all forest types with very low representation of other groups (Orthoptera and Blattaria). *Herbivores:* Herbivore guild was recorded from evergreen and montane forests and was the least abundant guild in Evergreen forest.

Discussion

Composition of the ground surface dwelling arthropods in the moist forests of south Western Ghats is typical of the fauna recorded in tropical moist forests in Australia, Neotropics, Eastern Africa, and northern region of moist south Western Ghats (Frith & Frith, 1990; Burgess *et al.*, 1999; Richardson *et al.*, 2005; Anu *et al.*, 2009). Acari, Collembola, Formicidae and Coleoptera accounted for more than 70% of total fauna in the Australian and eastern African regions (Greenslade & Greenslade, 1968; Plowman, 1979; Peterson & Luxton, 1982;

Holt, 1985; Burgess *et al.,* 1999). Acari, Formicidae, Collembola, Isoptera and Coleoptera were the dominant fraction of ground surface dwelling arthropods in the neotropical forests in Puerto Rico (Richardson *et al.,* 2005). Acari, insect larvae, Coleoptera, Formicidae and Collembola dominated moist forests in the southern region of the Western Ghats whereas Coleoptera, Acari, Hymenoptera, Collembola, insect larvae dominated the moist forests in the northern region of the south Western Ghats (Anu & Sabu, 2006; Anu *et al.,* 2009). Whereas Acari, insect larvae, Coleoptera, Formicidae and Collembola dominated the low and mid-elevation forests and Acari, insect larvae, Coleoptera, Collembola and Araneae in the high altitude montane shola forests in the southern region of the Western Ghats. Such variations indicate that relative abundance of ground dwelling arthropods vary between and within zoogeographical regions which might be influenced by the variations in topography, rainfall, litter moisture content, and litter depth in tropical forests (Olson, 1994; Bruhl *et al.,* 1999; Wagner *et al.,* 2003; Vineesh *et al.,* 2007) and also by problems of unequal sampling of some previous studies which makes generalizations difficult.

High relative abundance of predominantly detritivorous insect larvae in all forest types indicates its prominence as a major group in the forest floor of the moist Western Ghats. Their prominence is attributed to the presence of larval forms of many adult insects who spend their adult stage as predator/herbivore in the canopy in addition to the entirely ground dwelling terrestrial insects (Wolda 1988). Waterlogged substrates, humid, cold conditions in montane shola forests and tough pachyphyll leaves of montane forest plant species inhibit microbial and detritivore decomposition of leaf litter. Further, cool temperatures slow down important metabolic processes and retard or prevent the development of eggs or insect larvae (Brown 1973; Edwards 1974; Grubb 1977). Hence, it might be the cold wet habitat conditions and low resource quality of detritus in montane shola forests which lead to the low abundance of insect larvae. Aseasonal availability of leaf litter as breeding habitat to litter dwelling larval stages of arboreal insects and entirely terrestrial insects along with longer leaf life span of evergreen tree species and availability of leaves of different age classes in canopy to arboreal herbivores might have lead to the high abundance of insect larvae in evergreen forests. Low abundance of insect larvae in deciduous forests arises from the seasonal availability of thick litter cover both as a habitat and food resource for a short period following the presummer annual leaf shedding of deciduous trees and synchronization of the breeding and feeding activities of herbivorous and detritivorous insects with the phenology of deciduous trees species in the Western Ghats (Ananthakrishnan, 1996; Sabu & Vinod, 2009).

High proportional abundance of Collembola and trend towards higher abundance in montane shola forests results from their ability to persist in the cold wet climate of the montane shola forests. Moist habitats indirectly benefit Collembola by enhancing the growth of fungi, their main food source. Collembola migrate upwards in response to microorganismal blooms that occur after rainstorms (Hassall *et al.,* 1986; Fergusson & Joly, 2002; Lensing *et al.,* 2005). Hence, larger fungal populations in the moist montane shola forests, may have contributed to their higher abundance. Further, in addition to fungal resource availability, low abundance of Formicidae as a major predator (Olson, 1994; Sabu *et al.,* 2008) would have aided in their population build up as a prominent fraction in montane shola forests. Prominence of Coleoptera in montane shola forest arises from the high incidence of small sized humicolous Hydrophilidae; Leiodidae; Lathridiidae and Hydraenidae and herbivorous Curculionidae contrary to their lower representation in evergreen and deciduous forest types (unpublished data).

Comparable levels of abundance of Formicidae in both deciduous and evergreen forests agrees with the records from the north region of Western Ghats (Vineesh *et al.,* 2007; Sabu *et al.,* 2008), where contrary to the expected high abundance of Formicidae in the more open and dry deciduous forests (Bruhl *et al.,* 1999), no variations in the abundance betwen evergreen and deciduous forests was obvious. High abundance Formicidae in the wet and humid evergreen forests arises from the high abundance of prey resources such as Collembola, insect larvae and Isoptera (Samson *et al.,* 1997; Shattuck & Barnett, 2001). Low abundance of Formicidae in the cold and wet montane shola forests and their position as a minor group is in accordance with the earlier records from the high elevation forests in south east Asia, neotropical and northern region of south Western Ghats (Olson, 1994, Bruhl *et al.,* 1999; Sabu *et al.,* 2008).

Preference of Araneae towards cool wet moist conditions and the abundance of Collembola which is their major prey resource (Harwood *et al.,* 2001, 2003; Wise, 2004) lead to the high proportional abundance of Araneae in montane shola forests.

Another factor which might have contributed to the higher relative abundance of Araneae in montane litter floor is its ecological replacement of Formicidae and carabids as the prominent predator (Sabu *et al.,* 2008) contrary to the observations (Holldobler & Wilson, 1990, Olson, 1994) about the replacement of Formicidae by carabids and cicindelids.

Isoptera is a key detritivore groups in tropical ecosystems, playing pivotal roles in mediating the decomposition process and regulating nutrient and energy flows (Lawton *et al.,* 1996). Their cryptic nature, relative inactivity, range of nesting habits and feeding strategies by making underground tunnels, and patchy distributions within and between habitats makes sampling efficiency low leading to under presentation in terrestrial arthropod studies and it necessitates composite sampling essential for reasonable population assessments (Abensperg-Traun & De Boer, 1990, Eggleton *et al.,* 2002, Dawes-Gromadzki & Spain, 2003). Abysmally low abundance of Isoptera, in all forest vegetation types in the moist Western Ghats highlights the inefficacy of heat mediated Berlese sampling method in collecting ground surface foraging Isoptera and the need for composite sampling.

Predominance of ticks, as a prominent fraction in the montane shola forests is a pattern not recorded in other forest types. Association of ticks with deer and mountain goat, the prominent mammalian fauna in shola forest-grassland mosaics, and presence of small sized montane shola forest patches as the only shelter available for mammals in the midst of vast stretches of shola forest-grassland mosaics (ENP, 2009) might have contributed to their build up in montane shola forest. High proportional abundance of Psocoptera as a major fraction in deciduous forest alone indicates its preference towards more open and less wet forests. Low abundance of Hemiptera in montane shola forests is indicative of its low preference towards cold wet litter conditions. It is likely that Lepidoptera in the evergreen forest floor are accidental species (tourists) from the canopy or open areas which utilize litter as an occasional shelter.

Functional guild composition of litter fauna in the moist Western Ghats confirms what many others have previously shown (Harding & Stuttard, 1974, Swift *et al.,* 1979), that fungivores are the most abundant functional guild in terms of number of individuals. High densities of fungivores consisting mostly of Acari and Collembola are attributed to the amount and quality of detritus and microbial productivity (Seastedt 1984, Hattenschwiler *et al.,* 2005) in the Western Ghats. Predominance of Acari which comprises in the forest floor 50-65% of the overall faunal composition in all forest types accounts for the high abundance of fungivores in the region.

Detritivores was the second largest guild in all forest types with distinctly high abundance in evergreen forests. High proportional abundance of detritivores in evergreen forests arises from the higher abundance of insect larvae. Though of low resource quality, aseasonal presence of larval forms of litter dwelling insects and canopy insects related to the seasonality of leaf shedding and sprouting in evergreen forests (Lowman 1982, Santiago *et al.,* 2005, Wood *et al.,* 2005, Anu *et al.,* 2009) contribute to the abundance of detritivores in evergreen forests. Low abundance of detritivores in montane evergreen forests is attributable to the from low resource quality of detritus of cold resistant tough pachyphyll leaves of montane forest plant species and pressure cold wet habitat conditions, less suitable for larval development and foraging (Brown, 1973, Edwards, 1974, Grubb, 1977, Olson, 1994). Low abundance in deciduous forests arises from the seasonal availability of litter substratum to litter dwelling larval stages of arboreal groups and for the entirely litter dwelling detritivore forms to a brief period following annual leaf shedding and synchronous mass leaf sprouting of deciduous tree species (Sabu & Vinod, 2009).

Comparison of proportional abundance of omnivores and predators across the forest types in the moist Western Ghats reveals that low proportional abundance of omnivores and high proportional abundance of predatory elements is a characteristic feature of montane shola forests alone in the region. The most noticeable variation in the functional guild composition among different forest vegetation types was in the very low abundance of omnivores in montane shola forests arising from the low abundance of Formicidae for reasons already discussed. High proportional abundance of predators in montane shola forest arises from the high relative abundance of Araneae, ticks and Pseudoscorpions. Primary reason for abundance of predators in the montane shola forests might be the high relative abundance of Collembola, major prey source for Araneae, ground dwelling Formicidae and Pseudoscorpions (Moulder & Reichle, 1972, Chen & Wise, 1999, Harwood *et al.,* 2001, 2003, Wise, 2004). Low abundance of omnivores and the prominence of predators in montane shola forests indicate replacement of omnivores comprising predominantly of Formicidae by predators consisting of generalist feeders and omnivores (Scheu & Setälä, 2002) in montane shola forests. Ecological replacement of Formicidae by predatory groups in the montane shola forests (Olson, 1994, Sabu *et al.,* 2008) has broader implications leading to the ecological replacement of the guild of omnivores by predators.

Presence of herbivores as a prominent guild above omnivores in the litter floor of montane shola forests is contrary to the pattern in evergreen and deciduous forests. Higher representation of curculionid beetles (Sabu & Shiju (submitted)) lead to the high proportional dominance of herbivores and their ranking above omnivores in montane shola forests. Similar pattern of high incidence of Curculionidae surpassing Staphylinidae as prominent litter dwelling beetle is recorded from the montane forests in Neotropics (Anderson & Ashe, 2000, Jones *et al.,* 2008). Since curculionids are strictly phytophagous and usually found in association with their host plants and their association with epiphytes such as orchids, ferns, mosses, lichens and bryophytes that form a nearly continuous covering on the trunks of the trees from the ground to the tips of the highest branches in the montane shola forests (Nair and Menon, 2001, Anderson, 2004, Yanoviak *et al.,* 2004) indicates that it may be availability of leaf resources from the epiphytes that facilitates the abundance of Curculionidae in montane shola forests.

Sampling bias in multi-taxa sampling of ground surface dwelling arthropods

Behavioural differences among faunal elements and the characteristics of the habitat, where sampling is to be done, strongly influence the sampling techniques (Southwood, 1978, Melbourne, 1999, Woodcock, 2005). Hence, its possible that sources of error associated with sampling of mulit-taxa arthropods from the floor of moist deciduous, moderately wet evergreen and highly wet montane forests with spatially and temporarily restricted density-based quadrat sampling technique (Berlese funnel extraction method) might have failed to capture nocturnally active and crepuscular groups and the larger and more agile taxa (Robertson, 2007, Sabu & Shiju (in press)). Moreover chances for the death of heat sensitive groups such as Collembola and insect larvae and Isoptera with narrow ecological tolerance, before they drop into the collection jars is also equally high with Berlese extraction method (Besuchet *et al.,* 1987, Didham *et al.,* 1998). Hence, it is logical to judge that the sampling technique employed might have resulted in under representation of many groups and introduces a conservative bias in the empirical data. However, recent multivariate analyses indicated that morphospecies composition in samples was more strongly influenced by habitat type than by sampling methods (Missa *et al.,* 2007). This implies, for the purpose of sampling and comparing the diversity and abundance of muti-taxa ground dwelling arthropod assemblages in well contrasted forest types, differences in trap efficiency may be neglected and a employment of a single protocol that is less labor intensive, economical and that could capture many taxa is justifiable.

Acknowledgements

Financial assistance provided by the Department of Science and Technology (DST), Government of India is gratefully acknowledged.

References

Abensperg-Traun M, and De Boer ES. 1990. Species abundance and habitat differences in biomass of subterranean termites (Isoptera) in the wheat belt of Western Australia. *Australian Journal of Ecology* 15: 219–226.

Ananthakrishnan TN. 1996. *Forest Litter Insect Communities: Biology and Chemical Ecology.* New Delhi, Oxford & IBH.

Ananthakrishnan TN. 2000: Changing landscapes, ecodiversity and insect faunal dynamics: The Tenmalai (Western Ghats) experience: *Current Science* 78: (4) 25

Anderson RS, and Ashe JS. 2000. Leaf litter inhabiting beetles as surrogates for establishing priorities for conservation of selected tropical montane cloud forests in Honduras, Central America (Coleoptera; Staphylinidae, Curculionidae). *Biodiversity and Conservation* 9: 617–653.

Anderson RS. 2004. Family 131. Curculionidae, *In*: RH Arnett Jr., MC Thomas, PE Skelley & JH Frank (eds.), *American beetles.* CRC, New York. pp. 722–815.

Andresen E. 2002. Dung beetles in a Central Amazonian rain forest and their ecological role as secondary seed dispersers. *Ecological Entomology* 27: 257–270.

Anu A, and Sabu TK. 2007. Biodiversity analysis of forest litter ant assemblages in Wynad region of Western Ghats with taxonomic and conventional diversity measures. *Journal of Insect* Science 6:60, available online: http:// www.insectscience. org/6.60.

Anu A, Sabu TK, and Vineesh PJ. 2009. Seasonality of litter insects and relationship with rainfall in a wet evergreen forest in south Western Ghats. 10pp. *Journal of Insect Science* 9:49, available online: insectscience.org/9.49

Anu A, and Sabu TK. 2006. Litter arthropod diversity and community structure in an evergreen forest in the Wayanad region of Western Ghats. *Entomon* 31 (3): 181–190.

Besuchet C, Burckhardt DH, and Lobl I. 1987. The "Winkler/Moczarski" eclector as an efficient extractor for fungus and litter Coleoptera. *Coleopterists Bulletin* 41: 392-394.

Boinski S, and Fowler NL. 1989. Seasonal patterns in a tropical lowland forest. *Biotropica* 21(3): 223–233.

Borror DJ, De Long DM, and Triplehorn CA. 1996. *An introduction to the study of insects,* pp 873. Saunders.

Brown WL. 1973. A comparison of the Hylean and Congo-West African rainforest ant faunas. *In:* BJ Meggers, ES Ayensu & MID Duckworth (eds.), *Tropical forest ecosystems in Africa and South America: a comparative review.* Smithsonian Tropical Press, Washington DC, USA. pp. 161–185

Brühl CA, Mohamed M, and Linsenmair KE. 1999. Altitudinal distribution of leaf litter ant along a transect in primary forest on mount Kinabalu, Sabah, Malaysia. *Journal of Tropical Ecology* 16: 265–267.

Burgess ND, Ponder KL, and Goddard J. 1999. Surface and leaf-litter arthropods in the coastal forests of Tanzania. *African Journal of Ecology* 37(3): 355–365.

Chen B, and Wise DH. 1999. Bottom-up limitation of predaceous arthropods in a detritus-based terrestrial food web. *Ecology,* 80:761–772.

Clark DA, Piper SC, Keeling CD, and Clark DB. 2003. Tropical rain forest tree growth and atmospheric carbon dynamics linked to interannual temperature variation during 1984–2000. *Proceedings of the National Academy of Sciences USA* 100: 5852–5857.

Coll M, and Guershon M. 2002. Omnivory in terrestrial arthropods: mixing plant and prey diets. *Annual review of Entomology* 47: 267–297.

Dawes-Gromadzki TZ, and Spain A. 2003. Seasonal patterns in the activity and species richness of surface-foraging termites (Isoptera) at paper baits in a tropical Australian savanna. *Journal of Tropical Ecology*19: 449–456.

Didham RK, Hammond PM, Lawton JH, Eggleton P, and Stork NE. 1998. Beetle species responses to tropical forest fragmentation. *Ecological Monographs* 68: 295–323.

Edwards CA. 1974. Macroarthropods. *In*: CH Dickinson & GJF Pugh (eds.), *Biology of plant litter decomposition. Vol. 2.* Academic Press, London, England. pp. 523–554.

Eggleton P, Davies RG, Connetable S, Bignell DE and Rouland C. 2002. The termites of the Mayombe Forest Reserve, Congo (Brazzaville): transect sampling reveals an extremely high diversity of ground nesting soil feeders. *Journal of Natural History* 36: 1239–1246.

ENP 2009. Eravilkulam national park- Fauna and flora. Department of Forests and Wildlife, Government of Kerala. site assessed on 30th June 2009. *http://www.eravikulam.org/managementplan.html.*

Ferguson SH, and Joly DO. 2002. Dynamics of springtail and mite populations: the role of density dependence, predation, and weather. *Ecological Entomology* 27: 565–573.

Frith D, Frith C. 1990. Seasonality of Litter Invertebrate Populations in an Australian Upland Tropical Rain Forest. *Biotropica* 22(2): 181–190.

Giller PS. 1996. The diversity of soil communities, the 'poor man's tropical rainforest. *Biodiversity and Conservation* 5: 135–168.

Greenslade PJM, and Greenslade P. 1968. Soil and litter fauna densities in the Solomon Islands. *Pedobiologia* 7: 362–70.

Grubb PJ. 1977. Control of forest growth and distribution on wet tropical mountains: with special reference to mineral nutrition. *Annual Review of Ecology and Systematics* 8: 83–107.

Harding JL, and Stuttard RA. 1974. Microarthropods. *In*: CH Dickinson & GJE Pugh (eds.), *Biology of Plant Litter Decomposition. Vol. 2.* London: Academic press. pp. 489–532.

Harwood JD, Sunderland KD, and Symondson WOC. 2001. Living where the food is: web location by linyphiid spiders in relation to prey availability in winter wheat. *Journal of Applied Ecology, 38,* 88–99.

Harwood JD, and Sunderland KD, Symondson WOC. 2003. Web-location by linyphiid spiders: prey-specific aggregation and foraging strategies. *Journal of Animal Ecology* 72: 745–756.

Hassall M, Visser S, and Parkinson D. 1986. Vertical migration of *Onychiurus subtenuis* (Collembola) in relation to rainfall and microbial activity. *Pedobiologia* 29: 175–182.

Hättenschwiler S, Tiunov AV, and Scheu S. 2005. Biodiversity and litter decomposition in terrestrial ecosystems. *Annual Review of Ecology, Evolution & Systematics* 36: 191–218.

Hobbs RJ. 1994. Landscape ecology and conservation moving from description to application. Pacific Conservation Biology 1: 170–176.

Hölldobler B, and Wilson EO. 1990. *The ants.* Belknap Press of Harvard University. Press Cambridge.

Holt JA. 1985. Acari and Collembola in the litter and soil of three north Queensland rainforests. *Australian Journal of Ecology* 10: 57–65.

Hunt HW, Coleman DC, Ingham ER, Ingham RE, Elliott ET, Moore JC, Rose SL, Reid CPP, and Morley CR. 1987. The detrital food web in a shortgrass prairie. *Biology and Fertility of Soils* 3: 57–68.

Jones RW, O'brien CW, Ruiz-Montoya L, and Gomwz-Gomez B. 2008. Insect Diversity of Tropical Montane Forests: Diversity and Spatial Distribution of Weevils (Coleoptera: Curculionidae) Inhabiting Leaf Litter in Southern Mexico. *Annals of Entomological Society of America.* 101 (1): 128–139

Lawton JH, Bignell DE, Bloemers GF, Eggleton P, and Hodda ME. 1996. Carbon flux and diversity of nematodes and termites in Cameroon forest soils. *Biodiversity and Conservation* 5: 261–273.

Lee KE, and Pankhurst CE. 1992. Soil organisms and sustainable productivity. *Australian Journal of Soil Research* 30: 855–92.

Lensing JR, Todd S, and Wise DH. 2005. The impact of altered precipitation on spatial stratification and activity-densities of springtails (Collembola) and spiders (Araneae). *Ecological Entomology* 30 (2): 194–200.

Lowman MD. 1982. Seasonal variation in insect abundance among three Australian rainforests, with particular reference to phytophagous types. *Australian Journal of Ecology* 7: 353–361.

Melbourne BA. 1999. Bias in the effect of habitat structure on pitfall traps: an experimental evaluation. *Australian Journal of Ecology* 24: 228–239.

Missa O, Basset Y, Alonso A, Miller SE, Curletti G, Meyer MD, Eardley C, Mansell MW, and Wagner T. 2007. Monitoring arthropods in a tropical landscape: relative effects of sampling methods and habitat types on trap catches. *Journal of Insect Conservation* 1–18.

Moore IC, Walter DE, and Hunt HW. 1988. Arthropod regulation of micro- and mesobiota in below-ground detrital food webs. *Annual Review of Entomology* 33: 419–439.

Moulder BC, and Reichle DC. 1972. Significance of spider predation in the energy dynamics of forest floor arthropod communities. *Geological Monograph.* 42: 473–498.

Myers N, Mittermeier RA, Mittermeier CG, Da Fonseca GAB, and Kent J. 2000. Biodiversity hotspots for conservation priorities. *Nature* 403: 853–858.

Nair KKN, and Menon ARR. 2001. Endemic arborescent flora of the sholas of Kerala and its population and regeneration status . *In*: KKN Nair *et al.*, (eds.), *Shola forests of Kerala: Environment and biodiviersity.* Kerala forest Department and Kerala forest research Institute. pp. 209–233.

New TR. 1999.Descriptive taxonomy as a facilitating discipline in invertebrate conservation. *In:* W. Ponder, D. Lunney (eds.). *The other 99%.The conservation and biodiversity of invertebrates.* Transactions of the Royal Zoological Society of New South Wales, Mosman, Australia. pp.154–158.

Olson DM. 1994. The distribution of leaf litter invertebrates along a Neotropical altitudinal gradient. *Journal of Tropical Ecology* 10: 129–150.

Peterson H, and Luxton M. 1982. A comparative analysis of soil fauna populations and their role in decomposition process. *Oikos* 39: 287–388.

Plowman KP. 1979. Litter and soil fauna of two Australian subtropical forests. *Australian Journal of Ecology* 4: 87-104.

Richardson BA, Richardson MJ, Soto-Adames FN. 2005. Separating the effects of forest type and elevation on the diversity of litter invertebrate communities in a humid tropical forest in Puerto Rico. *Journal of Animal Ecology* 74: 926-936.

Robertson HG. 2007. Comparison of leaf litter ant communities in woodlands, lowland forests and montane forests of north-eastern Tanzania. *Biodiversity and Conservation* 11: 1637–1652.

Sabo JL, Soykan CU, and Keller A. 2005. Functional roles of leaf litter detritus in terrestrial food webs. *In:* PC De Ruiter, V Wolters & J Moore (eds.), Dynamic Food Webs: Multispecies assemblages, ecosystem development, and environmental change. Academic Press. pp. 223-234.

Sabu TK, and Shiju TR. (In Press). Efficacy of pitfall trapping, Winkler and Berlese extraction methods for measuring ground-dwelling arthropods in moist-deciduous forests in the Western Ghats. *Journal of Insect Science* online: *http://www. insectscience.org.*

Sabu TK, Vineesh PJ, and Vinod KV. 2008. Litter ant diversity along the altitudinal gradients of Wynad Forests. *Journal of Insect Science* 8:01, available online: *http://www. insectscience.org/8.01.*

Sabu TK, and Vinod KV. 2009. Population dynamics of the rubber plantation litter beetle Luprops tristis, in relation to annual cycle of foliage phenology of its host, the para rubber tree, *Hevea brasiliensis. Journal of Insect Science* 9: 41, available online: insectscience.org/9.41

Sabu TK. 2009. Litter insect diversity and trophic interactions in relation to the substrate biochemical profile in the major forest vegetations of Kerala. Project report submitted to Department of Science and Technology, Government of India.

Samson DA, Rickart EA, and Gonzales PC. 1997. Ant diversity and abundance along an elevational gradient in the Philippians. *Biotropica* 29: 349–363.

Santiago LS, Silvera K, and Schuur EAG. 2005. Nutrient cycling and plant-soil feedbacks along a precipitation gradient in lowland Panama. *Journal of Tropical Ecology,* 21:461-470.

Scheu S, and Setälä H. 2002. Multitrophic interactions in decomposer communities. *In:* T Tscharntke & BA Hawkins (eds.), Multitrophic Level Interactions. Cambridge University Press. pp. 223–64.

Seastedt TR. 1984. The role of microarthropods in decomposition and mineralization processes. *Annual Review of Entomology* 29: 25–46.

Shattuck SO, and Barnett NJ. 2001. Australian ants online, CSIRO, Australia. Available online: *www.ento.csiro.au/science/ants/formicinae.* Dated 10.07.2006

Southwood TRE. 1978. Ecological methods, with particular reference to the study of insect populations. 2nd edition, Chapman and Hall, London.

Swamy HR, and Proctor J. 1994. Litter fall and nutrient cycling in four rain-forests in the Sringeri area of the Indian Western Ghats. *Global Ecology and Biogeography* 4 (5): 155-165.

Swift MJ, Heal OW, and Anderson JM. 1979. *Decomposition in terrestrial ecosystems.* 1st Edition. University of Chicago Press, Berkeley and Los Angeles, 509 pp.

Vineesh PJ, Sabu TK, and Karmaly KA. 2007. Community structure and functional group classification of litter ants in the montane evergreen and deciduous forests of Wayanad region of Western Ghats, Southern India. *Oriental Insects* 41: 185-200.

Vulinec, K. 2002. Dung beetle communities and seed dispersal in primary forest and disturbed land in Amazonia. *Biotropica* 34: 297-309.

Wagner JDW, Toft S, and Wise DH. 2003. Spatial stratification in litter depth by forest-floor spiders. *Journal of Arachnology* 31: 28-39

Wise DH. 2004. Wandering spiders limit densities of a major microbi-detritivore in the forest-floor food web. *Pedobiologia* 48: 181–188.

Wolda H. 1988. Insect seasonality: Why? *Annual Review of Ecology and Systematics* 19: 1-18.

Wood TE, Lawrence D, and Clark DA. 2005. Variation in leaf litter nutrients of a Costa Rican rain forest is related to precipitation. *Biogeochemistry* 73: 417-437.

Woodcock BA. 2005. Sampling theory and practice. *In*: SR Leather (ed.), *Insect sampling in forest ecosystems.* Blackwell publishing. pp. 37–57.

Yanoviak SP, Walker H, and Nadkarni NM. 2004: Arthropod assemblages in vegetative vs. humic portions of epiphyte mats in a neotropical cloud forest. *Pedobiologia* 48 : 51–58.

Chapter - 3

Biodiversity of Insects in Spice Crop Ecosystems

S. Devasahayam and T.K. Jacob
Indian Institute of Spices Research, Marikunnu P.O., Calicut-673 012 (Ker.)

Introduction

Biodiversity can be defined as the variety of life on earth at all its levels, from genes to ecosystems, and the ecological and evolutionary processes that sustain it. Insects and related arthropods which are the most diverse group of organisms on earth, are an important component of biodiversity and make up more than 95% of the known animal diversity globally. Insects are also major contributors to biodiversity in agro-ecosystems which are the largest of ecosystems occupying almost 40% of the land area on earth. There are virtually no ecosystems in the world that are natural (having escaped human influence) and most ecosystems have been to some extent modified or cultivated by human activity for the production of food and livelihood security.

Insect biodiversity is important in agro-ecosystems not only for crop production through pollination, but also for other ecological services such as being a part of food chain, decomposition of organic matter, recycling of nutrients and suppression of undesirable organisms. Although crop pests are widely perceived as threats to agricultural productivity, their wider role in the ecosystem tends to be overlooked. It is estimated that the biodiversity of parasitoids and predators in these ecosystems provide 5-10 times more insect control than provided by insecticides worldwide. Agriculture is often viewed as the primary destructive force of biodiversity, but agro-ecosystems also hold the key to saving the richness of life on earth. Insect biodiversity is a sensitive indicator of health of various ecosystems including agro-ecosystems due to their abundance and rapid response to climate and other environmental changes. Insect biodiversity in these ecosystems is also influenced by other factors such as presence of other crops and weed plants and natural enemies in these ecosystems.

In India about 50% of the total land is cultivated making agro-ecosystems the largest of all ecosystems. Among the varied crops grown in these ecosystems, spice crops are grown in an area of 2.41 crore hectares in most of the states occupying about 12% of the cultivated area in the country. The products of these crops termed 'spices' add flavour and taste to food, besides being widely used in cosmetics and medicine. India is a leading producer, consumer and exporter of spices in the world. Spice crops are grown under varied ecological conditions ranging from humid tropical hills in south India to hot dry plains in north India and many of them are grown in hotspots of biodiversity such as western ghats. Although over 50 species of plants yield spices in India, black pepper (*Piper nigrum* L.), cardamom (*Elettaria cardamomum* Maton), ginger (*Zingiber officinale* Rosc.) and turmeric (*Curcuma longa* L.) account for a considerable share in export earnings for the country. Unlike other agricultural or horticultural crops, these four spice crops are grown in characteristic situations which include many other crops, support and shade trees and also weeds influencing the diversity and dynamics of interaction of insect pests along with their natural enemies, pollinators and other insects in these ecosystems. The biodiversity of insects in black pepper, cardamom, ginger and turmeric ecosystems and the various factors that influence the biodiversity in these ecosystems are discussed in this chapter.

Black pepper

Black pepper (*Piper nigrum* L.) (Piperaceae), is a perennial climbing vine mainly cultivated in Kerala and Karnataka from the plains to high altitudes up to about 1500 m above MSL (Fig. 1a). The spice of commerce is obtained from the mature dried berries. The humid tropical evergreen forests of western ghats of south India is the centre of origin and biodiversity of black pepper. Black pepper is trailed on suitable supports termed as standards and the most commonly used standards in plantations are *Erythrina* spp., silver oak (*Grevillea robusta* A. Cunn. ex R.Br.), *Glyricidia sepium* (Jack.) Kunth ex Walp., *Ailanthes* sp. and *Garuga pinnata* Roxb. In homestead gardens, black pepper is trailed on other economically important trees such as coconut

(*Cocos nucifera* L.), arecanut (*Areca catechu* L.), jack (*Artocarpus* spp.), mango (*Mangifera indica* L.), etc.

Ninety six species of insects belonging to 9 orders and 35 families have been recorded to be associated with black pepper in India, which includes 64 species of insect pests feeding on the plant, 29 species of natural enemies and 5 species are ants attending mealybug colonies and *successori* in leaf thrips galls (Fig. 2 a-c; Fig. 3 a-d). Among them, the Order Hemiptera comprising of 10 families is the most dominant group forming of 39.5% of the insects associated with black pepper (Table 1). Among the various insect pests, *pollu* beetle (*Lanka ramakrishnai* Prathapan & Viraktamath), scale insects (*Lepidosaphes piperis* Gr. and *Aspidiotus destructor* Sign.), top shoot borer (*Cydia hemidoxa* Meyr.), leaf gall thrips (*Liothrips karnyi* Bagn.) and root mealybugs (*Planococcus* spp.) are important (Devasahayam 2000a).

Table 1. Biodiversity of insect species on black pepper

Order	Insect pests		Natural enemies		Total	
	No. of families	No. of species	No. of families	No. of species	No. of families	No. of species
Orthoptera	1	1	-	-	1	1
Phasmida	1	1	-	-	1	1
Hemiptera	9	36	1	2	10	38
Thysanoptera	1	3	1	3	1	6
Neuroptera	-	-	1	1	1	1
Coleoptera	4	13	2	9	5	22
Diptera	1	3	3	3	3	6
Hymenoptera	1*	5*	6	10	6	15
Lepidoptera	6	7	1	1	7	8

* Includes ants attending mealybug colonies and *successori* in leaf thrip galls

The *pollu* beetle is the most destructive insect pest of black pepper especially in plains and midlands in Kerala. The adults feed and damage tender shoots, leaves and spikes and the larvae bore into and feed on internal tissues of tender spikes and berries. Very few natural enemies have been recorded on *pollu* beetle and the only species which occasionally preys on *pollu* beetle is *Oecophylla smargdina* (Fab.) (Formicidae) (Devasahayam & Koya 1994).

Among the various species of scale insects recorded on black pepper, mussel scale and coconut scale are serious pests at higher altitudes especially in Idukki, Wayanad (Kerala), Kodagu (Karnataka) and Nilgiris (Tamil Nadu) districts (Koya *et al.*, 1996). These two species encrust stems, leaves, and berries and feed on plant sap resulting in chlorotic patches on leaves, wilting and drying of branches and mortality of younger vines. A majority of the natural enemies (17 species) have been documented on scale insects which account for 58.6% of the natural enemies recorded on insect pests of black pepper. Among them, *Aphytis* sp. (Aphelinidae), *Pseudoscymnus* sp. and *Chilocorus circumdatus* Gyllen. (Coccinellidae) are more common being widely distributed in black pepper plantations at higher altitudes (Selvakumaran *et al.*, 1996).

The top shoot borer is a serious pest of younger vines in all black pepper areas and the larvae bore into tender terminal shoots resulting in their decay and affect the growth of the vine. Seven species of insect natural enemies have been recorded on the top shoot borer among which five belong to the Order Hymenoptera. At Peruvannamuzhi (Kerala), hymenopterous parasitoids (among which *Apanteles cypris* Nixon (Braconidae) was the most common) parasitized up to 20% of larvae/pupae during September to November (Devasahayam & Koya 1993).

Leaf gall thrips are serious pests at higher altitudes especially on younger vines and also on cuttings in the nursery. The pest infestation causes crinkling and malformation of leaves and affects the growth of younger vines and cuttings. Five species of insect natural enemies have been documented on leaf gall thrips among which *Montandoniola moraguesi* Puton (Heteroptera) and *Androthrips flavipes* Schmutz (Phlaeothripidae) are widely distributed in all black pepper areas and are voracious predators feeding on all stages of leaf gall thrips (Devasahayam, 2000b). A unique feature of leaf thrip galls is the occurrence of a rich fauna of arthropods within the galls which includes 35 species among which 21 species are insects. These comprise of primary consumers of gall tissues, natural enemies of primary consumers, detritus feeders and *successori* which colonized the galls after it was free of primary consumers and natural enemies. There was also a definite pattern of succession of the arthropod fauna in the thrip galls (Devasahayam 2000b).

Root mealybugs infest roots of black pepper vines especially at higher altitudes resulting in yellowing, wilting and drying of leaves and mortality of young vines. Except for larvae of *Spalgis* sp. (Papilionidae) no other natural enemy has been recorded on these hypogeic insect pests. Five species (unidentified) of ants were commonly associated with root mealybug colonies especially in Wayanad and Kodagu districts (Devasahayam 2006).

Cardamom

Cardamom (*Elettaria cardamomum* Maton) (Zingiberaceae), is a perennial herb mainly cultivated under the shade of forest trees in the hilly tracts of Kerala, Karnataka and Tamil Nadu (Fig. 1b). The spice of commerce is obtained from the dried fruit or capsules. The evergreen forests of the western ghats of south India is the centre of origin of cardamom. Cardamom is generally cultivated under natural forest canopy except in certain areas in Uttara Kannada, Chickamagalur and Hasan in Karnataka and Wayanad in Kerala where it is grown as a intercrop in arecanut or coffee gardens.

Ninety one species of insects belonging to 8 orders and 42 families have been recorded to be associated with cardamom in India which includes insect pests (54 species), natural enemies (34 species) and pollinators of cardamom flowers (3 species) (Fig. 2 d). Among them, the Order Lepidoptera and Hymenoptera are the most dominant groups forming of 26.4% each of the insects recorded on cardamom (Table 2).

Table 2. Biodiversity of insect species on cardamom

Order	Insect pests		Natural enemies		Total	
	No. of families	No. of species	No. of families	No. of species	No. of families	No. of species
Orthoptera	1	2	-	-	1	2
Hemiptera	9	15	1	1	10	16
Thysanoptera	2	2	-	-	2	2
Neuroptera	-	-	1	2	1	2
Coleoptera	4	8	1	3	4	11
Diptera	2	3	2	7	4	10
Hymenoptera	1*	3*	4	21	5	24
Lepidoptera	15	24	-		15	24

* Includes pollinators of cardamom flowers

Among the insect pests infesting cardamom, cardamom thrips (*Sciothrips cardamomi* Ramk.), shoot and capsule borer (*Conogethes punctiferalis* Guen.) and root grub (*Basilepta fulvicorne* Jacoby) are important (Table 2). In addition, nine species of hairy caterpillars are reported to damage cardamom among which *Eupterote cardamomi* Renga., *E. canarica* Moore, *E. undata* Blanchard and *Lenodora vittata* Wlk. are important (Gopakumar & Chandrasekhar 2002).

Cardamom thrips is the most widespread and destructive pest of cardamom damaging shoots, panicles and capsules and resulting in shedding of flowers and immature capsules and formation of scabs, splitting and shrivelling of mature capsules. Very few natural enemies have been recorded on thrips and they include two species of general predators such as anthocorids and chrysopids (Varadarasan 1995).

The shoot and capsule borer is a serious pest in the nursery and plantation. The larvae bore into unopened leaves, panicles, immature capsules and shoots and feed on the internal tissues. Eighteen species of hymenopterous parasitoids have been recorded on the shoot and capsule borer among which *Eriborus trochanteratus* Morl. (Ichneumonidae) and *Xanthopimpla australis* Kr. (Ichneumonidae) are more common parasitizing 10.5%-11.1% of larvae at Appangala (Karnataka) (CPCRI 1985). *Friona* sp. and *Agrypon* sp. parasitized 8%-10% and 16%-18% of larvae/pupae, respectively, at Idukki (Kerala) (Varadarasan 1995).

The root grub is a serious pest in the nursery and plantation and the larvae feed on roots and portions of rhizomes leading to yellowing and stunting in growth of infested clumps. No natural enemy (insect) has been recorded on the root grub.

Hairy caterpillars occur sporadically during certain years in large numbers and defoliate cardamom plants completely. Severe defoliation retards plant growth and adversely affects yield; the pest attack on nursery plants is often fatal. A single species of dipteran and hymneopteran parasitoid have been recoded on hairy caterpillars (Varadarasan 1995).

Recent studies conducted by Mathew (2008) indicate that 190 species of insects belonging to 10 orders and 53 families occur in the cardamom and forest tree ecosystem in Nelliampathy (Kerala). The fauna in the cardamom area was characterized by herbaceous feeding species with a wide host range and adapted to feed on a variety of secondary tree species.

Fig. 1. Ecosystems of major spice crops (a) Black pepper mixed cropped with coffee at Wayanad District, Kerala (b) Cardamom under forest trees at Kodagu District, Karnataka (c) Ginger intercropped with coffee and black pepper at Wayanad District, Kerala (d) Turmeric intercropped with coconut at Kozhikode District, Kerala.

Ginger and turmeric

Ginger (*Zingiber officinale* Rosc.) and turmeric (*Curcuma longa* L.) are perennial rhizomatous herbs grown in many states in India under various agro-climatic conditions (Fig. 1c & d). Ginger and turmeric are believed to have originated in south East Asia but have been under cultivation since ancient times in India. Ginger is grown predominantly in Kerala, Meghalaya and Arunachal Pradesh whereas, Tamil Nadu, Orissa and Andhra Pradesh, are predominant turmeric growing states.

Thirty one species of insects have been recorded on ginger among which 21 species are insect pests and saprophytic feeders and 10 species are natural enemies. Among them, the Order Diptera is the most dominant group constituting 38.7% of the total species (Table 3). Ninety one species of insects have been recorded on turmeric among which 64 species are insect pests and saprophytic feeders, and 29 species natural enemies (Fig 2e

& f; Fig. 3 e & f). Among them, the order Coleoptera is the most dominant group constituting 25.2% of the total species (Table 4).

Table 3. Biodiversity of insect species on ginger

Order	Insect pests		Natural enemies		Total	
	No. of families	No. of species	No. of families	No. of species	No. of families	No. of species
Isoptera	1	1	-	-	1	1
Hemiptera	3	3	-	-	3	3
Thysanoptera	1	1	-	-	1	1
Coleoptera	2	4	-	-	2	4
Diptera	6*	9*	2	3	8	12
Lepidoptera	3	3	-	-	3	3
Hymenoptera	-	-	5	6	5	6
Dermaptera	-	-	1	1	1	1

* Includes saprophytic feeders

Table 4. Biodiversity of insect species on turmeric

Order	Insect pests		Natural enemies		Total	
	No. of families	No. of species	No. of families	No. of species	No. of families	No. of species
Orthoptera	2	4	-	-	2	4
Hemiptera	9	15	-	-	9	15
Thysanoptera	1	1	-	-	1	1
Neuroptera	-	-	1	1	1	1
Coleoptera	4	19	1	4	5	23
Diptera	6*	11*	2	3	8	14
Lepidoptera	7	11	-	-	7	11
Dermaptera	-	-	1	2	1	2
Odonata	-	-	1	1	1	1
Hymenoptera	-	-	9	19	9	19

* Includes saprophytic feeders

Among the various insect pests, the shoot borer (*Conogethes punctiferalis* Guen.) is the most serious and widespread pest of ginger and turmeric and the larvae bore into shoots and feed on the central shoot affecting the yield of rhizomes (Devasahayam & Koya 2004; 2007). A majority of natural enemies have been documented on shoot borer infesting ginger and turmeric which account for 61.8% and 73.9%, respectively, of the natural enemies recorded on these crops. These include dermapterans, asislid flies and hymenopterous parasitoids. At Peruvannamuzhi (Kerala), *Apanteles taragammae* and *Bracon* sp. (Braconidae) were the common natural enemies of shoot borer and were active in the field during September to December (Devasahayam 1996). Six species of dipteran flies belonging to the families Platystomatidae, Muscidae, Chloropidae and Milichidae have been recorded to be associated with shoot borer affected turmeric plants in Kerala feeding on the rotting tissues around the site of damage caused by the shoot borer (Jacob, 1981). The rhizome scale (*Aspidiella hartii* Sign.) is a serious insect pest during storage of ginger and turmeric rhizomes; however, the pest infestation also occurs during later stages of the crop in the field. Three species of hymenopterous parasitoids have been recorded on the pest (Koya *et al.,* 1991).

Among the other insect pests, the leaf roller (*Udaspes folus* (Cram.) is common during the monsoon season in Kerala and Karnataka and six species of natural enemies belonging to Diptera and Hymenoptera have been recorded on the pest (CPCRI 1977; Sujatha *et al.,* 1992). Six species of natural enemies belonging to Odonata, Dermaptera and Hymenoptera have been recorded on the rhizome fly (*Mimegralla coeruleiformis* Macq.) that is considered as an important pest of ginger and turmeric in Karnataka, Andhra Pradesh and Maharashtra (Koya 1989; Kotikal & Kulkarni 2000). However, the pest is generally seen in ginger and turmeric rhizomes infected with soft rot disease and is probably a secondary saprophytic feeder.

Influence of multiple cropping on biodiversity

The perennial nature of many of the spice crops and the availability of other crops, shade and support trees along with various insect species have a profound influence on the biodiversity of insects in these ecosystems. Black pepper is cultivated in multiple cropping systems in many plantations especially at higher altitudes along with other crops such as coffee (*Coffea* spp.), tea (*Camelia sinensis* (L.) Kuntze), cardamom (*Eletaria cardamomum* Maton), cocoa (*Theobromae cacao* L.), *Citrus* sp. and vanilla (*Vanilla planifolia* Andrews), whereas in homesteads in the plains and midlands it is cultivated along with ginger (*Zingiber officinale* Roscoe), turmeric (*Curcuma longa* L.), taro (*Colacasia* spp.), tapioca (*Manihot esculenta* Crantz), banana (*Musa* sp.), etc.

Cardamom is generally cultivated with coffee and black pepper especially in Kodagu and Idukki districts. Trees belonging to over 32 families of angiosperms constitute the major tree flora in cardamom hills in South India (Shankar 1980). Recent studies conducted by Mathew (2008) indicated that 42 species of plants occur in cardamom plantations in Nelliampathy (Kerala) which included 27 species of trees and the remaining shrubs and herbs.

Ginger is grown as a pure crop or as an intercrop in coconut, arecanut and coffee gardens. It also forms a component of homestead farming and is grown with a variety of crops such as vegetables (beans, cucumber, lady's finger, chillies and cabbage), pulses (pigeon pea and black gram), cereals (maize and finger millet), oil seeds (castor, soybean, sunflower) and tubers (tapioca, taro and yams) in many regions. Turmeric is grown as an intercrop in coconut and arecanut gardens in Kerala and Karnataka and is raised as a mixed crop with vegetables (chillies, onion and brinjal), pulses (pigeon pea and green gram) and cereals (maize and ragi) in many other states.

Influence of alternate hosts on biodiversity

Spice crops such as black pepper, are characterized by distinctive insects of monophagous nature, whereas in other spice crops such as cardamom, ginger and turmeric, generalist feeders are widely distributed. Most of the major insect pests of black pepper such as *L. ramakrishnai*, *L. piperis*, *C. hemidoxa* and *L. karnyi* are either monophagus or have a very narrow host range. Other insect pests such as the root mealy bugs (*Planococcus* spp.) and *A. destructor* have a wide host range occurring on many economically important crops. *Planococcus* spp. occur on black pepper and other economically important crops such as coffee, banana, cardamom, ginger and turmeric during the monsoon and post-monsoon seasons, but generally shift to weed plants during the summer season (Devasahayam 2006).

A few insect pests of cardamom such as *C. punctiferalis* and *S. cardamomi* have a wide host range. *C. punctiferalis* has been recorded on more than 65 host plants in India and some of the zingiberaceous species like *Amomum subulatum* Roxb., *A. microstephanum* Baker, *Aframomum melegueta* Schum., *Hedychium coronarium* J. Konig, *H. flavescens* Carey ex Rosc., *Alpinia galanga* (L.) Sw., *Z. officinale* and *C. longa* are commonly observed in the vicinity of cardamom plantations (Nambiar *et al.,* 1975; CPCRI 1985).

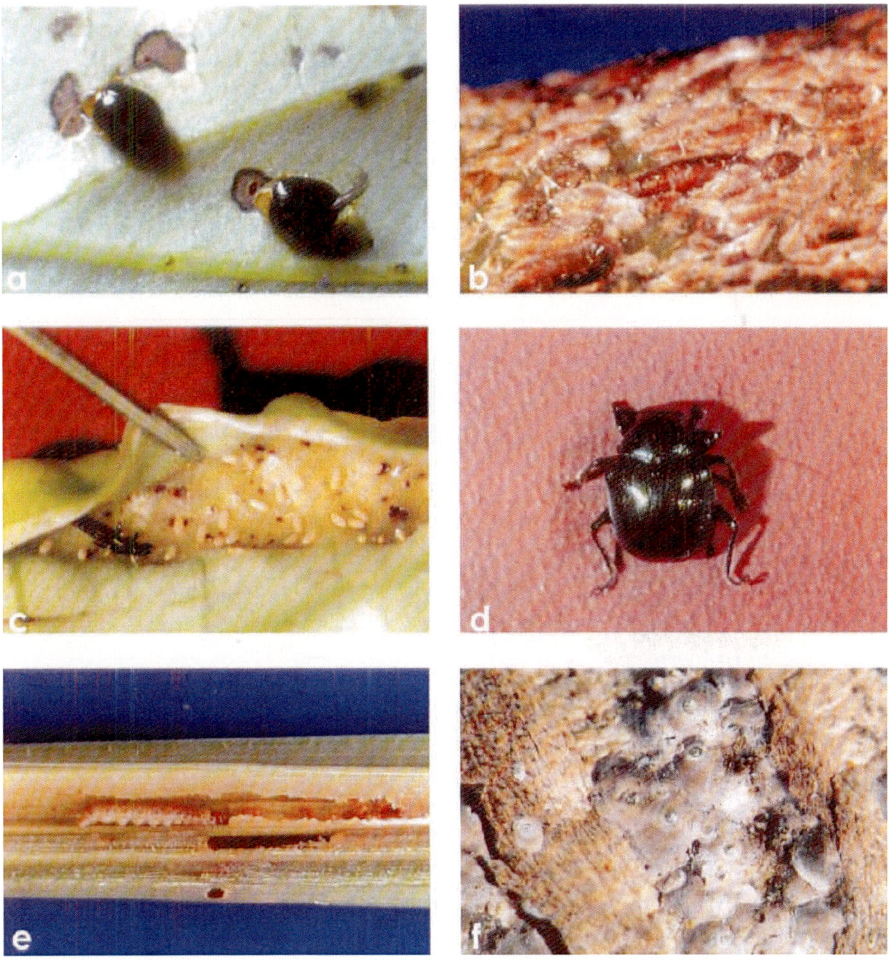

Fig. 2. Biodiversity of insect pests of major spice crops (a) Pollu beetle (*Lanka ramakrishnai*) infesting black pepper (b) Mussel scale (*Lepidosaphes piperis*) infesting black pepper (c) Leaf gall thrips (*Liothrips karnyi*) infesting black pepper (d) Adult of root grub (*Basilepta fulvicorne*) infesting cardamom (e) Larvae of shoot borer (*Conogethes punctiferalis*) infesting turmeric (f) Rhizome scale (*Aspidiella hartii*) infesting ginger.

The cardamom thrips feeds and breeds on over 10 species of zingiberaceous and graminaeceous plants such as *A. cannae-carpum* Benth., *A. involucratum* Benth., *Aframomum melegueta* (Rosc.), *Hedychium flavescens* Carey, *H. coronarium* Koen., *Alpinia* sp., *A galanga* Sw., *Zingiber* sp., *Z. officinale* Rosc., *Z. casumunar* Roxb., *Z. nimmonii* Dalz., *Z. wightianum* Thw., *Curcuma longa* Linn., *C. pseudomontana* Grah., *C. zeodaria* Rosc., *Panicum longipes* W. et A., *Remusatia vivipara* Schott,

Colocasia antiquorum Schott, *Musa* sp., *Costus speciosus* Sm., *Crinum* sp. and *Globba ophioglossa* Wt. that grow wild in and around cardamom plantations (Nambiar *et al.*, 1975; Kumaresan *et al.*, 1988).

Fig. 3. Biodiversity of natural enemies of insect pests of major spice crops (a) *Montandoniola moraguesi* predating on larvae of leaf gall thrips (b) *Androthrips flavipes* predating on eggs of leaf gall thrips (c) Larvae of *Chilocorus circumdatus* predating on scale insects (d) Larvae of *Pharoscymnus* sp. predating on scale insects (e) *Apanteles* sp. parasitic on shoot borer larvae (f) *Cocobius* sp. parasitic on rhizome scale.

The cardamom aphid (*Pentalonia nigronervosa* Van der Goot) also feed and breed on other hosts belonging to Araceae such as *Colacasia* sp., *Alocasia* sp. and *Caladium* sp. that are common in the cardamom plantations. The hairy caterpillars (*Eupterote* spp., *Linodera vittata* Walk., *Euproctis lutifacia* Hamp. and *Alphaea biguttata* Walk. and *Pericallia ricini* Fabr.) are polyphagous feeding voraciously on the forest trees during their early stages and shift to cardamom during later stages. The adults of root grubs feed on trees such as jack (*Artocarpus heterophylla*), Indian almomd (*Terminalia catapha* L.), mango (*Mangifera indica* L.), guava (*Psidium guajava* L.), ficus (*Ficus indica* Hochst. ex Walp., *F. bengalensis* L.), cocoa (*T. cacao*) and *Erythrina* spp. that are common around cardamom plantations (Varadarasan & Naidu 1993).

The leaf roller infesting ginger and turmeric has also been observed on arrowroot, *Curcuma augustifolia* Roxb., *E. cardamomum*, *A. melegueta*, *C. amada* and *Hedychium* sp (Koya *et al.*, 1991). Maggots of the rhizome fly *M. coeruleifrons* also feed on rhizome rot affected turmeric, wild arrow root, *Colacasia* sp., wild ginger (*Zingiber* sp.), *C. aromatica*, *C. zeodaria* and *K. galanga* (Koya 1989).

Changes in biodiversity

Climatic changes in the form of lower rainfall and higher minimum and maximum temperatures during the past decades has resulted in changes in insect biodiversity of certain species in spice crop ecosystems. Insect pests such as *L. ramakrishnae* that were not observed at higher altitudes in Kerala and Karnataka earlier are becoming increasingly important at Wayanad, Idukki and Kodagu districts. The whitefly *Kanakarajiella cardamomi* (David & Subramanian) which was a minor pest on cardamom is now serious in certain pockets in Idukki District and Palani Hills. Invasive pests such as the *Erythrina* gall wasp (*Quadrasticus erythrinae* Kim) have also recently established in black pepper ecosystems in south India where *Erythrina* spp. are grown as standards and also for shade. The phenomenon of 'host range expansion' has also been observed in the root mealybugs *Planococcus* spp. which were major insect pests on coffee especia-lly in Wayanad and Kodagu districts earlier but are now economically important on black pepper during the past two decades.

Economic development and other factors associated with rapid increases in population, which accelerate habitat conversion and environmental degradation affect insect biodiversity in spice ecosystems also as in any other ecosystems. However, a fundamental problem in estimating biodiversity loss is that very little quantitative data are available for insects in these ecosystems. Though the species diversity of insects in various spice crops and other intercrops has been estimated individually, no information is available on the species richness in these ecosystems.

References

Central Plantation Crops Research Institute (CPCRI). 1977. Annual Report for 1976, Kasaragod, p. 283.

Central Plantation Crops Research Institute (CPCRI). 1985. Annual Report for 1983, Kasaragod, p. 249.

Devasahayam, S. 1996. Biological control of insect pests of spices. In: Anandaraj, M. & Peter, K.V. (Eds.) Biological Control in Spices, Indian Institute of Spices Research, Calicut. pp. 33–45.

Devasahayam, S. 2000a. Insect pests of black pepper. In: Ravindran, P. N. (Ed) Black Pepper *Piper nigrum*. Harwood Academic Publishers, Amsterdam. pp. 309–334 .

Devasahayam, S. 2000b. Bioecology of leaf gall thrips *Liothrips karnyi* Bagnall infesting black pepper. PhD Thesis, Submitted to University of Calicut, Calicut, p. 199.

Devasahayam, S. 2004. Insect pests of ginger. In : Ravindran, P. N. & Babu, K. N. (Eds.) Ginger. The Genus *Zingiber*. CRC Press, Washington. pp. 367–389.

Devasahayam, S. 2006. Bio-ecology and integrated management of root mealybug (*Planococcus* sp.) infesting black pepper. Final Report of Ad-hoc Research Scheme. Indian Institute of Spices Research, Calicut. pp. 26.

Devasahayam, S. & Koya, K.M.A. 1993. Seasonal incidence of hymenopterous parasites of top shoot borer *Cydia hemidoxa* (Lepidoptera : Tortricidae) on black pepper. J. Entomol. Res. 17 : 205–208.

Devasahayam, S. & Koya, K.M.A. 1994. Natural enemies of major insect pests of black pepper (*Piper nigrum* L) in India. J. Spices Arom. Crops 3 : 50–55.

Devasahayam, S. & Koya, K.M.A. 2007. Insect pests of turmeric. In: Ravindran, P.N., Babu, K.N. and Sivaraman, K. (Eds.) Turmeric. The Genus *Curcuma*. CRC Press, Boca Raton. pp. 169–192.

Gopakumar, B. & Chandrasekhar, S. S. 2002. Insect pests of cardamom. In : Ravindran, P.N. and Madhusoodanan, K.J. (Eds.). Cardamom. The Genus *Elettaria*. Taylor and Francis, London. pp. 180–206.

Jacob, S.A. 1981. Biology of *Dichocrocis punctiferalis* Guen. on turmeric. J. Plantn. Crops 9: 19-23.

Koya, K.M.A. 1989. Bio-ecology of *Mimegralla coeruleifrons* Maquart (Diptera: Micropezidae) associated with ginger *Zingiber officinale* Rosc. rhizomes. Entomon 14: 81-84.

Koya, K.M.A., Devasahayam, S. & Premkumar, T. 1991. Insect pests of ginger and turmeric in India-a review. J. Plantn. Crops 19 : 1–13.

Koya, K.M.A., Devasahayam, S., Selvakumaran, S. & Kallil, M. 1996. Distribution and damage caused by scale insects and mealybugs associated with black pepper (*Piper nigrum* Linn.) in India. J. Entomol. Res. 20: 129–136.

Kotikal, Y.K. & Kulkarni, K.A. 2000. Insect pests infesting turmeric in Northern Karnataka. Karnataka J. Agric. Sci. 13 : 858–866.

Kumaresan, D., Regupathy, A. & Bhaskaran, P. 1988. Pests of Spices. Rajalakshmi Publications, Nagercoil, 241 pp.

Mathew, G. 2008. Plant and insect biodiversity patterns in cardamom plantations vis-a-vis natural forests: an assessment in the Nelliyampathy Hills, Kerala, India. In: Kannaiyan, S. & Gopalam, A. (Eds.) Agrobiodiversity Volume III. Diversity in Pests and Pathogens. Associated Publishing Company, New Delhi, pp. 69-75.

Nambiar, M.C., Pillai, G.B. & Nambiar, K.K.N. 1975. Diseases and pests of cardamom-a reume of research in India. Pesticides Annual 1975, pp. 122-127.

Selvakumaran, S., M. Kallil and S. Devasahayam 1996. Natural enemies of two major species of scale insects infesting black pepper (*Piper nigrum* L.) in India. Pest Manage. Hort. Ecosyst. 2, 79–83.

Shankar, B.D. 1980. Shade trees and their role in cardamom cultivation. Cardamom 12 (3): 3–11.

Sujatha, A., Zaherudeen, S.M. and Reddy, R.V.S.K. 1992. Turmeric leaf roller, *Udaspes folus* Cram. and its parasitoids in Godavari Delta. Indian Cocoa Arecanut Spices J. 15: 118–119.

Varadarasan, S. 1995. Biological control of insect pests of cardamom. In : Ananthakrishnan, T.N. (Ed.) *Biological Control of Social Forest and Plantation Crops Insects.* Oxford & IBH Publishing Company Private Limited, New Delhi. pp. 109–119.

Varadarasan, S. and Naidu, R. 1993. Studies on Root Grubs. Consolidated Report on Research Project Financed by NABARD (1989–93). Indian Cardamom Research Institute, Myladumpara. 55 p.

Chapter - 4

Insect biodiversity in beverage crops Ecosystems

N. Muraleedharan

UPASI Tea Research Foundation, Valparai 642 127, Coimbatore (TN) India
Email: advisor.upasitearesearch@yahoo.com

The major beverage crops grown in India are tea and coffee and to a minor extent, Cocoa. In fact cocoa plants are grown as an intercrop in homestead gardens and has not acquired the status of a major commercial crop in India. Tea is cultivated in an area of more than 5,21,000 ha in the northern, north eastern and southern parts of India. The commercially cultivated tea plants are derived from *Camellia sinensis* (L) O. Kuntze, the short leaved 'China' plants, *Camellia assamica* (Masters) Wight, the broad leaved 'Assam' plants and *C. assamica lasiocalyx* (Planchon ex Watt) Wight the 'Cambod' plants, a subspecies of the latter and their numerous hybrids. Tea prefers a warm humid climate, well distributed rainfall and long sunshine hours. The plant exhibits alternate phases of active growth and dormancy and is highly sensitive to photoperiod. Dormancy sets in when day length remains below 11 hours 15 minutes for about six weeks and this happens in winter at latitudes beyond about 16°N (Baruah, 1989). However, in south India where tea is grown at latitudes between 8° and 12°N, the bushes do not experience winter dormancy and they flush throughout the year. In India the tea crop is commercially grown as large plantations and the tea ecosystem is unique since it is grown as a monoculture under shade trees (Fig. 1). Tea plantations provide excellent habitat for several arthropod species, due to their perennial nature, continuous supply of food and offering suitable sites for reproduction. Another significant feature of the tea plantations is their close proximity to the forests and in certain cases to other crops like coffee, oranges, vegetables, cardamom, pepper, cloves and rice.

Tea plants are grown at different elevations, ranging from sea level to higher elevations, up to 2400 m above mean sea level. The temperature in these areas also exhibit wide variations, from sub zero to 38°C. In winter months, the plants at higher elevations suffer from frost damage and in summer the temperature may go up 38°C, especially in the tea plantations of north east. The annual rainfall, again could widely vary from 80 cm to 500 cm. Obviously, the climate and physical features of the tea growing areas differ considerably, offering a wide choice for the colonization and diverse insect species.

More than one thousand species of insects are associated with tea as pests, casual visitors and also as predators and parasitoids of pests (Muraleedharan and Chen, 1997). Besides, there are the insects associated with the shade trees grown in the plantations. The dynamic adaptation of insects have enabled them to occupy every part of the tea plant and the maximum number of arthropods occur on the leaves (Table 1). Among them Lepidoptera is the largest containing 32% of the species, followed by Hemiptera with 27% of the species. The foliage, stems, shoots, flowers and fruits offer habitat for several species, though the flowers and fruits are not of much economic significance in tea. The foliage feeders are more abundant than others and these include the insects belonging to Lepidoptera, Thysanoptera and Hemiptera, besides a complex of acarine species. The branches are frequented by many species of stem boring and bark eating caterpillars, scolytid beetles and termites. The roots of tea plants are eaten by grubs of scarabaeid beetles and termites which include both, live wood and scavenging species. Occasionally, root infesting scale insects and mealy bugs invade not only the nursery plants but also the roots of mature tea bushes.

Table 1. Number of insect and mite pests recorded on different parts of the tea plant (Chen & Chen, 1989)

Plant part	No. of pests	percentage of total number of pests
Foliage	734	58.9
Stem	300	24.1
Root	160	12.8
Flowers	30	2.4
Seeds	14	1.1

Each geographic region has its own distinctive pest fauna on tea, though several species have been found from more than one region. The number of insects associated with the tea plants depends on the length of time for which it is cultivated in that region. The area under tea became important only after allowing for the "age effect" and latitude has no influence on species richness. In large tea areas, saturation in species richness reached over a period of 100-150 years. The north Asian countries with the longest period of tea growing have the largest number of insect and mite species while the south American countries where tea was rather recently introduced have very few pests. The accumulation of arthropod species on tea, a perennial plant, is also influenced by the age of plants, the plantations with old tea harbour more number of insect species. (Banerjee, 1983).

Tea in north east India is grown under a complex of shade tree species which help to reduce the high temperature experienced in the summer months and also contribute to the conservation of soil moisture, recycling of soil nutrients and reduction in tea leaf temperature *Albizzia chinensis* is the most commonly planted shade tree in north east India and the other commonly planted trees are *Albizzia. odoratissima, A.procera* and *Dalbergia assamica.* Also grown are *Albizzia lebbek, Acacia lenticularis, Derris robusta* and *Dalbergia sissoo.* As a temporary shade tree, *Indigofera teysmanii* has been proved to be very successful. In south, only one species, *Grevillea robusta,* the silver oak is planted as a shade tree in the plantations, mainly because it gives filtered light and does not compete with the tea plants for nutrients and moisture. This exotic tree has no serious pests and diseases.

The shade trees are frequented by a number of insect species, such as foliage feeders, bark eaters, pinhole borers, shot hole borers and heart wood borers. About 150 species are known as pests of these shade trees and green crops in north east India (Table 2).

Table 2. Insects affecting shade trees (Ray Chaudhuri and Saha, 2000)

Astycus chrysochlorus wied	Curculionidae.Col
Holotricia impressa Burn	Scarabaeidae. Col
Diacrissia obliqua Walker	Arctiidae. Lep
Dichomeris ianthes Meyr	Gelechiidae .Lep
Buzura suppressaria Guenee	Geometridae . Lep
Pericyma umbrina Guenee	Noctuidae.Lep
Rhesela moestalis Walker	Noctuidae.Lep
Eurema hecabe (L)	Pieridae. Lep

Eurema blanda silhetana (Wallace)	Pieridae.Lep
Boradesa omissa (Roths & Jord.)	Sphingidae. Lep
Agrillus beesoni Obenberger	Buprestidae. Colloptera
Cryptorrhynchus sp.	Curculionidae. Col
Leptaulax dentatus (Fabr.)	Passalidae. Col
Inderbela quadrinotata (Walker)	Inderbellidae. Lep
Psylla oblonga Mathur	Psyllidae.Homoptera
Trichilizum bisulcatus Fabricius	Brenthidae. Colloptera
Phoephilus agrestis Boheman	Bostrychidae
Belionota prasina Thunberg	Buprestidae.Col
Aeolesthes sp.	Cerambycidae.Col
Exocentrus fumosus Gahan	Cerambycidae.Col
Hoplocerambyx spinicornis Newman	Cerambycidae.Col
Pachydissus parvicollis Gahan	Cerambycidae.Col
Stromatium barbatum (Fabricius)	Cerambycidae.Col
Xylotrechus smei Lap & Gory	Cerambycidae.Col
Xystrocera globosa Oliver	Cerambycidae.Col
Oreoderus bhutanus Arrow	Scarabaeidae.Col
Trichochrysea hirta (Fabricius)	Membracidae. Hom
Sternocera aurosignata Thompson	Buprestidae.Col

The presence of insects feeding on the tea plants and shade tress invariably lead to the buildup of several species of insect parasitoids and predators (Muraleedharan *et al.,* 1988, Muraleedharn *et al.,* 2001) (Table 3). In fact, the minor status of several pests is due to action of these natural enemies. Two such cases of excellent natural regulation are of the scale insect *Fiorinia theae* and the aphid *Toxoptera aurantii.* (Nagarkatti *et al.,* 1979, Radhakrishnan *et al.,* 1988, Radhakrishnan and Muraleedharan, 1993). The tea plantations harbour several species of coccinellids such as *Cryptogonus bimaculatus, Jauravia quandrinotata, J. opaca, Menochilus sexmaculatus* and *Stethorus gilvifrons,* which feed on eriophyids, spider mites, aphids and scale insects. Several syrphid flies are active in the tea fields and these include *Episyrhus balteatus, Paragus tibialis, Betasyrphus serarius, Ischiodon scutellaris* and *Allobaccha nubilipennis,* primarily feeding on the aphids. The hemerobiid, *Micromus timidus* has a wide distribution in tea fields when compared to Chrysopids. The parasitoids, *Apanteles coedicius, Trioxys indicus* and *Aphidius colemani* have several hosts like *Toxoptera aurantii* feeding on tea and *Aphis gossipii* and *Aphis craccivora* frequenting the weeds in tea fields.

Table 3. Parasitoids and predators of tea pests in India

Natural enemies	Order and Family	Affected pests
+*Scolothrips rhagebianus* Priesner	Thysanoptera:Thripidae	Mites
+*Aeolothrips intermedius* Bagnall	Thysanoptera: Aeolothripidae	Thrips
+*Mymarothrips garuda* (Ramak.& Marg.)	Thysanoptera: Aeolothripidae	Thrips
+*Aspectus indicus* Arrow	Coleoptera: Dermestidae	Mites
+*Orphinus fucundus* Arrow	Coleoptera: Dermestidae	Mites
+*Caelophora* sp.	Coleoptera: Coccinellidae	Mites
+*Cryptogonus bimaculatus*	Coleoptera: Coccinellidae	Mites, aphids
+*Cryptogonus 4-guttatus* Weise	Coleoptera: Coccinellidae	Mites
+*Jauravia* sp.	Coleoptera:Coccinellidae	Mites
+*Jauravia soror* Weise	Coleoptera: Coccinellidae	Mites
+*Jauravia opaca* (Weise)	Coleoptera: Coccinellidae	Mites
+*Jauravia quandrinotata* Kapur	Coleoptera: Coccinellidae	Mites, aphids
+*Menochilus sexmaculatus* (F)	Coleoptera: Coccinellidae	Mites, aphids
+*Scymnus* sp.	Coleoptera: Coccinellidae	Mites, scale insects, aphids
+*Scymnus nubilus* Mulsant	Coleoptera: Coccinellidae	Mites
+*Stethorus gilvifron* Mulsant	Coleoptera: Coccinellidae	Mites
+*Stictobura* sp.	Coleoptera: Coccinellidae	Mites
+*Verania vincta* (Gorham)	Coleoptera: Coccinellidae	Mites
+*Jauravia pubescens* (Fab)	Coleoptera: Coccinellidae	Flushworm, aphids
+*Menochilus sexmaculatus*(Fab)	Coleoptera: Coccinellidae	Flushworm
+*Jauravia quadrinotata* Kapur	Coleoptera: Coccinellidae	Scale insects
+*Coccinella septumpunctata* L.	Coleoptera: Coccinellidae	Scale insects
+*Chilocorus nigiritus* (F.)	Coleoptera: Coccinellidae	Scale insects
+*Chilocorus circumdatus* Sch.	Coleoptera: Coccinellidae	Scale insects
+*Coleophora biplagiata Schw.* var. *melanota* Muls.	Coleoptera: Coccinellidae	Scale insects

+*Pharoscymnus horni* (Weise)	Coleoptera: Coccinellidae	Scale insects
+*Coccinella repanda* Thumb	Coleoptera: Coccinellidae	Aphids
+*Coccinella septempunctata* L.	Coleoptera: Coccinellidae	Aphids
+*Coccinella septempunctata* L. var. *diviaricata*	Coleoptera: Coccinellidae	Aphids
+*Coccinella transversalis* Fab.	Coleoptera: Coccinellidae	Aphids
+*Coleophora biplagiata* Schw.	Coleoptera: Coccinellidae	Aphids
+*Cryptogonus orbiculus* (Gyllenhal)	Coleoptera: Coccinellidae	Aphids
+*Leis dimidiata* (F.)	Coleoptera: Coccinellidae	Aphids
+*Leis dimidiata* (F) var, +*quindecimmaculata* Hope	Coleoptera: Coccinellidae	Aphids
+*Lemnia bissellata* (Mulsant)	Coleoptera: Coccinellidae	Aphids
+*Pseudaspidimerus circumflexus* (Motschulsky)	Coleoptera: Coccinellidae	Aphids
+*Scymnus pyrocheilus* Mulsant	Coleoptera: Coccinellidae	Aphids
+*Verania vincta* Gorh.	Coleoptera: Coccinellidae	Aphids
+*Oligota* sp.	Coleoptera: Staphylinidae	Mites
+*Calleida nilgirensis* Straneo	Coleoptera: Carabidae	Flushworm
+*Cybocephalus* sp.	Coleoptera: Nitidulidae	Mites
+*Chrysopa madestes* Banks	Neuroptera: Chrysopidae	Mites
+*Chrysoperla* sp.	Neuroptera: Chrysopidae	Aphids
+*Micromus timidus* Hegan	Neuroptera: Hemerobiidae	Aphids, flushworm, mites
+*Semidalis* sp. *fletcheri* Withycombe	Neuroptera: Coniopterygidae	Mites
+*Orius* sp.	Hemiptera: Anthocoridae	Mites, Thrips
+*Anthocoris* sp.	Hemiptera: Antocoridae	Thrips
+*Opistoplatys* sp.	Hemiptera: Reduviidae	Flushworm
+*Pseudopachybrachius guttus* (Dallas)	Hemiptera: Lygaeidae	Flushworm
+*Cantheconidea furcellata* (Wolff)	Hemiptera: Pentatomidae	Leaf roller, Bag worms, Nettle grubs
+*Triommata* sp.	Diptera: Cecidomyiidae	Scale insects
+*Microdon bellus* Brun	Diptera: Syrphidae	Scale insects

+*Allobaccha nubilipennis* (Austen)	Diptera : Syrphidae	Aphids
+*Betasyrphus adligatus* (Wiedemann)	Diptera : Syrphidae	Aphids
+*Betasyrphus serarius* (Wiedemann)	Diptera : Syrphidae	Aphids
+*Dideopsis aegrota* (Fab.)	Diptera : Syrphidae	Aphids
+*Episyrphus balteatus* (De Geer)	Diptera : Syrphidae	Aphids
+*lschiodon scutellaris* (Fab.)	Diptera : Syrphidae	Aphids
+*Paragus atratus* De Mejere	Diptera : Syrphidae	Aphids
+*Paragus indicus* Bron	Diptera : Syrphidae	Aphids
+*Paragus serratus* (Fab.)	Diptera : Syrphidae	Aphids
+*Paragus tibialis* (Fallen)	Diptera : Syrphidae	Aphids
+*Paragus yerburiensis* Stuckenberg	Diptera : Syrphidae	Aphids
+*Spalgis epeus* Westwood	Lepidoptera: Lycaenidae	Scale insects
* *Antherigona orientalis* Schin	Diptera: Muscidae	Red coffee borer
* *Sarcophaga* sp.	Diptera: Sarcophagidae	Looper caterpillar
* *Palexorista solennis* (Walker)	Diptera: Tachinidae	Tea tortrix
* *Nealsomyia rufella* Bezzi	Diptera: Tachinidae	Faggot worms
* *Blepharella* sp.	Diptera: Tachinidae	Cutworms
* *Carcelia* sp.	Diptera: Tachinidae	Cutworms
* *Peribaea orbata* (Wiesmann)	Diptera: Tachinidae	Cutworms
* *Winthemia* sp.	Diptera: Tachinidae	Cutworms
* *Cylindromya* sp.	Diptera: Tachinidae	Bunch Caterpillar
* *Peribaea* sp.	Diptera: Tachinidae	Red slug caterpillar, Nettle bug
* *Austrophorocera grandis* Macquart	Diptera: Tachinidae	Nettle grubs
* *Compsilura cocinnata*	Diptera: Tachinidae	Nettle grubs
* *Cocophagus cowperi* Girault	Hymenoptera: Aphelinidae	Brown scale
* *Aphytis* sp. nr. *mazalae* De bach & Rosen	Hymenoptera: Aphelinidae	
* *Aphytis theae* Cam.	Hymenoptera: Aphelinidae	Green scale
* *Aphytis fioriniae*	Hymenoptera: Aphelinidae	Green scale
* *Aspidiotiphagus* sp.	Hymenoptera: Aphelinidae	Green scale
* *Coccophagus bivittatus* Comp.	Hymenoptera: Aphelinidae	Green scale

* *Prospaltella* sp.	Hymenoptera: Aphelinidae	Green scale
* *Coccophagus acanthosceles* Wtstn.	Hymenoptera: Aphelinidae	Black scale
* *Aphytis* sp. *Chrysomphali* (Merc)	Hymenoptera: Aphelinidae	Red scale
* *Aphytis* sp.	Hymenoptera: Aphelinidae	Red scale
* *Aphterencyrtus microphagus* (Meyr.)	Hymenoptera: Aphelinidae	Red scale
* *Aphytis* sp. *chrysomphali* (Mercet)	Hymenoptera: Aphelinidae	Red scale
* *Aspidiotiphagus* sp.	Hymenoptera: Aphelinidae	Red scale
* *Aspidiotiphagus citrinus* Craw	Hymenoptera: Aphelinidae	Red scale
* *Coccophagus tachiachii* Madhihassan	Hymenoptera: Aphelinidae	Wax scale
* *Aphelinus* sp.	Hymenoptera: Aphelinidae	Aphids
* *Aphidius colemani* Vierick	Hymenoptera: Aphidiidae	Aphids
* *Lipolexis scutellaris* Mackauer	Hymenoptera: Aphidiidae	Aphids
* *Trioxys* sp.	Hymenoptera: Aphidiidae	Aphids
* *Trioxys indicus* Subba Rao and Sharma	Hymenoptera: Aphidiidae	Aphids
* *Goniozus* sp.	Hymenoptera: Bethylidae	Tea tortrix
* *Apanteles* sp.	Hymenoptera: Braconidae	Flushworm, leaf roller
* *Apanteles aristaeus* Nixon	Hymenoptera: Braconidae	Flushworm
* *Apanteles* sp. (*ater* group)	Hymenoptera: Braconidae	Flushworm, Tea tortrix
* *Apanteles coedicius* Nixon	Hymenoptera: Braconidae	Flushworm, Tea tortrix
* *Apanteles* sp. (*glomeratus* group)	Hymenoptera: Braconidae	Flushworm, Tea tortrix
* *Apanteles* nr. *malevolus* Wilkinson	Hymenoptera: Braconidae	Flushworm
* *Apanteles* sp. (*octonarius* group)	Hymenoptera: Braconidae	Flushworm
* *Asogaster* sp.	Hymenoptera: Braconidae	Flushworm, Tea tortrix
* *Bracon* sp.	Hymenoptera: Braconidae	Flushworm
* *Bracon* sp. nr. *greeni* Ashmead	Hymenoptera: Braconidae	Flushworm
* *Bracon* sp. nr. *hebetor* Say	Hymenoptera: Braconidae	Flushworm
* *Fornicia* sp.	Hymenoptera: Braconidae	Flushworm

* *Rogas* sp.	Hymenoptera: Braconidae	Flushworm
* *Apanteles fabiae* Wilkinson	Hymenoptera: Braconidae	Looper caterpillar
* *Apanteles taprobanae* Cameron	Hymenoptera: Braconidae	Lopper caterpillar, Red slug caterpillar
* *Apanteles* sp. (*ultor* group)	Hymenoptera: Braconidae	Large jelly grub
* *Batotheca nigriceps* Cam.	Hymenoptera: Braconidae	Common jelly grub
* *Spinaria spinator* Guer	Hymenoptera: Braconidae	Nettle grubs
* *Opius* sp.	Hymenoptera: Braconidae	Leaf miner
* *Aphanogmus manilae* (Ashmead)	Hymenoptera: Ceraphronidae	Looper caterpillar
* *Brachymeria excarinata* Gahan	Hymenoptera: Chalcidae	Tea tortrix
* *Brachymeria lasus* (Walker)	Hymenoptera: Chalcidae	Tea tortrix
* *Elasmus homonae* (Fer.)	Hymenoptera: Chalicidae	Tea tortrix
* *Elasmus homonae* (Ferr.)	Hymenoptera: Elasmidae	Flushworm
* *Elasmus krishnagirensis* Mani & Saraswat	Hymenoptera: Elasmidae	Leaf roller
* *Encrytus infelix* Embleton	Hymenoptera: Encyrtidae	Brown scale
* *Blastothrix* sp.	Hymenoptera: Encyrtidae	Black scale
* *Comperiella bifasciata* Howard	Hymenoptera: Encyrtidae	Black scale
* *Microterys* sp.	Hymenoptera: Encyrtidae	Black scale
* *Comperiella bifasciata* Howard	Hymenoptera: Encyrtidae	Red scale
* *Comperiella bifasciata* Howard	Hymenoptera: Encyrtidae	Red scale
* *Prospaltella* sp.	Hymenoptera: Encyrtidae	Red scale
* *Anagyrus dactylopii* (How)	Hymenoptera: Encyrtidae	Shoot mealy bug
* *Leptomastix nigrocoxalis* Compere	Hymenoptera: Encyrtidae	Shoot mealy bug
* *Asympiesiella india* Gir.	Hymenoptera: Eulophidae	Flushworm
* *Pediobius* sp.	Hymenoptera: Eulophidae	Flushworm
* *Tetrastichus* sp. (*miser*-group)	Hymenoptera: Eulophidae	Flushworm
* *Asympiesiella* sp.	Hymenoptera: Eulophidae	Leaf roller
* *Asympiesiella india* (Girault)	Hymenoptera: Eulophidae	Leaf roller
* *Elachertus* sp.	Hymenoptera: Eulophidae	Leaf roller
* *Mestocharella javensis* Gahan	Hymenoptera: Eulophidae	Leaf roller

* *Sympiesis dolichogaster* Ashmed	Hymenoptera: Eulophidae	Leaf roller
* *Sympiesis india* (Girault)	Hymenoptera; Eulophidae	Leaf roller
* *Euplectromorpha* sp. nr. *salomonis* Ferr.	Hymenoptera: Eulophidae	Nettle grubs
* *Euplectromorpha* sp. nr. *viridiceps* Ferr.	Hymoneptera: Eulophidae	Nettle grubs
* *Metaplectrus thoseae* Ferr.	Hymenoptera: Eulophidae	Nettle grubs
* *Aprostocetus purpureus* (Cameron)	Hymenoptera: Eulophidae	Shoot mealy bug
* *Metaplectrus thoseae* (Ferr.)	Hymenoptera: Eulophidae	Nettle grubs
* *Closterocerus insignis* Waters ton	Hymenoptera: Eulophidae	Leaf miner
* *Eurytoma* sp.	Hymenoptera: Eurytomidae	Flushworm, Tea tortrix
* *Aptesis* sp.	Hymenoptera: Ichneumonidae	Flushworm
* *Diadegma* sp.	Hymenoptera: Ichneumonidae	Flushworm
* *Eriborus* sp.	Hymenoptera: Ichneumonidae	Flushworm
* *Goniozus* sp.	Hymenoptera; Ichneumonidae	Flushworm
* *Meloboris* sp.	Hymenoptera: Ichneumonidae	Flushworm
* *Plectochorus* nr. *iwatensis* Uchida	Hymenoptera: Ichneumonidae	Flushworm
* *Pristomerus* sp.	Hymenoptera: Ichneumonidae	Flushworm
* *Diadegma* sp.	Hymenoptera: Ichneumonidae	Leaf roller
* *Genotropsis* sp.	Hymenoptera: Ichneumonidae	Leaf roller
* *Trathala* sp.	Hymenoptera; Ichneumonidae	Leaf roller
* *Apophua* sp.	Hymenoptera:Ichneumonidae	Tea tortrix
* *Phytodietus spinipes* (Cameron)	Hymenoptera: Ichneumonidae	Tea tortrix
* *Triclistus* sp.	Hymenoptera: Ichneumonidae	Tea tortrix
* *Erythmelus helopeltidis* Gahan	Hymenoptera: Mymaridae	Helopeltis
* *Oxyharma* sp.	Hymenoptera: Pteromalidae	Leaf miner
* *Trigonogastra* sp.	Hymenoptera: Pteromalidae	Leaf miner
* *Telenomus* sp.	Hymenoptera: Scelionidae	Tea tortrix
* *Microphanurus* sp.	Hymenoptera: Scelionidae	Tea seed bug

+ Predator; * Parasitoid

In North-East India the tea mosquito, *Helopeltis theivora* is preyed upon by *Chrysoperla carnea*, the coccinillids *Oxyopes* sp. *Plexippus* sp. *Phidippus* sp. *Scymnnus* sp. and mantids. A drynid wasp is an important parasitoid of the cicadellid *Empoasca flavescens*. The major parasitoids of caterpillar pests affecting tea are *Apanteles aristaeus, A. taprobanea, Sympiesis dolichogaster* and *Mestocharella javensis*. The intensity of activities of predators and parasitoids have been found to be very high in north east India (Gurusubramaniam *et al.,* 2008).

The Braconid, *Cotesia* sp and the tachinid, *Argyrophylax* sp. are very active in the tea areas in the foot hills of Darjeeling, causing 40-45% parasilization of the geometrid looper *Buzura suppressaria*. Parasitisation by *Argyrophylax* could even go up to 80% in the organically managed estates. The tea tortrix, *Homona coffearia* is under natural regulation by a series of parasitoids, dominated by the ichneumonid, *Phytodietus spinipes* (Muraleedharan and Selvasundaram, 1991). Among the predators active in the tea fields coccinellids and spiders are the most dominant. Braconids, ichneumonids and tachinids are the most common parasitoids (Rahman *et al.,* 2005). In the tea areas of north Bengal, Roy *et al.* (2005) found 35 species of spiders and 25 species of lady bird beetles.

The coffee plantations have certain close similarities to the tea ecosystem. Coffee cultivation is mostly confined to south India though to a limited extent it is grown in other parts of the country. Nearly 340,000 ha is under coffee crop. The commercially cultivated coffee plants belong to *Coffea arabica,* the arabica plant and *C.canephora,* the robusta plant. A third species *C. liberica,* the tree coffee is also grown on the estates along the hedges. Unlike tea, coffee is grown in large number of small holdings and the large growers are rather few. Coffee holdings with its densy canopy of shade trees help to preserve the forest ecosystem in the traditional coffee growing areas (Fig 2). They are located mostly in the hilly tracts of Western and Eastern ghats. Most of these areas receive the south west monsoon though a large part of the coffee tracts lie in the north east monsoon zone. Though coffee seedlings had been grown in backyards and gardens as early as in 1800, it was only in the middle of the century coffee became a plantation crop. By 1860, coffee cultivation reached its zenith and also witnessed the outbreak of major pests such as *Xylotrechus quadripes,* the white stem borer, followed by the green scale, *Coccus viridis.* Along with the leaf rust, fungus, *Hemileia vastatrix,* the two major pests caused serious havoc in the coffee plantations (Anon, 2003).

Although coffee is cultivated under shade, the sunlight intensity and temperature expressed during the drought period have been found to adversely affect the normal and healthy growth of coffee plants. It is a shade loving plant and its performance is better under shade. Therefore, maintaining adequate shade is of considerable significance for maintaining the physiological balance in coffee plants (Ramaiah & Raman, 1967) Shade trees help in maintaining an optimum light intensity and temperature, besides retaining the soil moisture and protecting the coffee plants from the effect of direct solar exposure since the light requirement of coffee plants is about 70%. Shade trees play a major role in the management of the dreaded white stem borer. The species composition of shade trees is rather diverse. The lower canopy consists of dadap *(Erythrina lithosperma)* which is lopped two or three times in an year, during the southwest monsoon. Next to dadap, silver oak, *Grevillea robusta* is the common temporary shade tree. The permanent shade trees are *Artocarpus integrifolia, Pterocarpus marsupium, Bischofia javanica, Dalbergia latifolia, Syzygium jambolana, Albizzia lebbeck, Albizzia stipulate, Albizzia odoratissima, Albizzia moluccana, Albizzia sumatrana, Ficus tsiela, Ficus tjakela, Ficus infectoria, Ficus glomerata, Ficus nervosa, Ficus retusa, Cedrella toona* and *Chloroxylon swietenia.*

The coffee ecosystem differs from the tea ecosystem, not only in having a denser shade but also because of the diversity of plants within the coffee fields or crops grown adjacent to the coffee estates. The coffee growing areas are also ideally suited for growing other crops such as pepper, cardamom, arecanut and citrus. Among these, pepper and orange trees are grown within the coffee fields while cardamom and arecanut are cultivated within the estate. The shade trees also serve as wind breaks and the pepper vines are trained on the shade trees. The arthropod diversity is much richer in such ecosystems, characterised by plant diversity.

All parts of the coffee plants are frequented by insects. More than a hundred species of insects have been recorded as pests of coffee in India (Table 5). The major pests are the white stem borer (*Xylotrechus quadripes*), affecting the main trunk, the shot hole borer (*Xylosandrus compactus*) boring in to the small branches, the berry borer, *Hypothenemus hampei,* leaf infesting green scale, *Coccus viridis* and the mealy bugs *Planococcus citri* and *P. lilacinus.* Besides these several other insects invade coffee plants (Somasekhar, 1958). Populations of *Coccus viridis* prefer higher temperature and low humidity for their development. Consequently, its population is higher on coffee plants exposed to sunlight rather than on plants growing under shade. Interestingly, when the scales are infected by the fungal pathogen *Cephalosporium lecanii,* there was greater mortality of those on shaded plants, than on the plants growing in the open (Easwaramoorthy and Jayaraj, 1997). The associated ant species such as *Crematogaster* and *Oecophylla smaragdina* afforded greater protection to the scales from natural enemies. The cultivation of coffee under greatly reduced shade has led to the outbreak of several pests.

Table 4. Insects affecting coffee plants

Insect pests	Family & Order	Affected plant parts
Xylotrechus quadripes	Cerambycidae. Col.	Stem Arabica
Hypothenemus hampei	Scolytidae. Col.	small branches (Robusta)
Xylosandrus compactus	Scolytidae. Col.	Berries
Zeuzera coffeae	Cossidae. Lep.	Stems
Planococcus citri	Zeuzeridae. Lep.	Leaves and tender stems
Planococcus lilacinus	Pseudococcidae. Hem.	Leaves and tender stems
Ferrisia virgata	Pseudococcidae. Hem.	Leaves and tender stems
Coccus viridis	Coccidae. Hem.	Leaves and tender stems
Saissetia coffeae	Coccidae. Hem.	Leaves and tender stems
Holotrichia spp.	Melolonthidae. Col.	Roots and leaves
Eupterote spp.	Bombycidae. Lep.	Leaves
Heliothrips haemorrhoidalis	Thripidae. Thy.	Leaves
Retithrips syriacus	Thripidae. Thy.	Leaves
Scirtothrips bispinosus	Thripidae. Thy.	Leaves
Homona coffearia	Tortricidae. Lep.	Leaves
Spodoptera litura	Noctuidae. Lep.	Leaves
Sahyadrassus malabaricus	Hepialidae. Lep.	Stems
Cerococcus ornatus	Astesolecaniidae. Hem.	Leaves and stems
Chloropulvinaria psidii	Coccidae. Hem.	Leaves
Tropicomyia sp.	Agromyzidae. Dip.	Leaf

Xyleborus globus	Scolytidae Col.	Stems
Xyleborus interjenctus	Scolytidae Col.	Stems
Xyleborus nitidus	Scolytidae Col.	Stems
Xyleborus noxius	Scolytidae Col.	Stems
Xyleborus conidens	Scolytidae Col.	Stems
Xylosandrus crassiusculus	Scolytidae Col.	Stems

Table. Natural enemies of coffee pests in India

Natural enemies	Order and Family	Affected life stages of pests
+*Allorbogas pallidiceps*		White stem borer
+*Apenesia* sp.	Hymenoptera: Bethylidae	White stem borer
+*Avetianella* sp.	Hymenoptera: Encyrtidae	White stem borer
+*Campyloneurus* sp.	Hymenoptera: Braconidae	White stem borer
+*Doryctes compactus*	Hymenoptera: Braconidae	White stem borer
+*Doryctes coxalis*	Hymenoptera: Braconidae	White stem borer
+*Gasteruption* sp.	Hymenoptera: Gasteroptionidae	White stem borer
+*Metapelma* sp.	Hymenoptera: Eupelmidae	White stem borer
+*Scleroderma vigilans*	Hymenoptera: Betrylidae	White stem borer
+*Scleroderma* sp.	Hymenoptera: Betrylidae	White stem borer
+*Euryoma* sp.	Hymenoptera: Eurytomidae	Short hole borer
+*Pyemotes berfsi*	Acasi: pyemotidae	Short hole borer
+*Tetrastichus* sp.nr *Xylebororum*	Hymenoptera: Eulophidae	Short hole borer
+*Callimerus* sp.	Coleoptera: Cleridae	Short hole borer
+*Cleridae* sp.	Coleoptera: Cleridae	Short hole borer
+*Eupelmus* sp.	Hymenoptera: Eupelmidae	Short hole borer
+*Formicidae* sp.	Hymenoptera: Formicidae	Short hole borer
+*Asynapta* sp.	Diptera: Coccinellidae	Short hole borer
+*Alemella flava*	Neuroptera: Coccinellidae	Mealy bugs
* *Aprostocerus purpureus*	Hymenoptera: Eulophidae	Mealy bugs
* *Anagyrus inopus*	Hymenoptera: Encyrtidae	Mealy bugs
* *Blepyrus insularis*	Hymenoptera: Encyrtidae	Mealy bugs
* *Homalotylus? indicus*	Hymenoptera: Encyrtidae	Mealy bugs
* *Leptomastix*	Hymenoptera: Encyrtidae	Mealy bugs
* *Nigrocoxalis*	Hymenoptera: Encyrtidae	Mealy bugs

* *Prochiloneurus* sp.	Hymenoptera: Encyrtidae	Mealy bugs
+*Dicrodiplosis* sp.	Diptora: Cecidomyiidae	Mealy bugs
+*Pseudoscymnus* sp.	Coleoptera: Coccinellidae	Mealy bugs
+*Pullus pallidicollis*	Coleoptera: Coccinellidae	Mealy bugs
+*spalgis epius*	Lepidoptera: Lycaenidae	Mealy bugs
+*Telsimia* sp.	Coleoptera: Coccinellidae	Mealy bugs
+*Triommata*	Diptera: Cecidomyiidiae	Mealy bugs
+*Coccidivora*	Hymenoptera: Chalcidae	Mealy bugs
* *Anagyrus* sp.	Hymenoptera: Braconidae	Mealy bugs
* *Apanteles nr. saurous*	Hymenoptera: Eulophidae	Mealy bugs
* *Aprostocerus?*	Diptera: Celyphidae	Mealy bugs
* *Purpureus*	Hymenoptera: Mymaridae	Mealy bugs
* *Gonatocerus* sp.	Hymenoptera: Aphidiidae	Mealy bugs
* *Gyranusoidea* nr. *flava*	Lepidoptera: Noctuidae	Mealy bugs
+*Eublema* sp.	Lepidoptera: Coccinellidae	Mealy bugs
+*Dicrodiplosis* sp.	Coleoptera: Coccinellidae	Mealy bugs
+*Hyperaspis*	Coleoptera: Coccinellidae	Mealy bugs
+*Maindroni*	Coleoptera: Histeridae	Mealy bugs
+*Leucopis luteicornis*	Hymenoptera; Ichneumonidae	Mealy bugs
+*Pallus pallidicollis*	Coleoptera: Coccinellidae	Mealy bugs
* *Sarcophaga (Boettcherisca) Peregrina* Scoliid wasps	Diptera: Sarcophagidae	Cockchafers, white grubs
* *Anagyrus qadrii*	Hymenoptera: Encyrtidae	Tailed mealy bug
* *Anicetus annulatus*	Hymenoptera: Encyrtidae	Tailed mealy bug
* *Blepyrus insularis*	Hymenoptera: Encyrtidae	Tailed mealy bug
+*Allogropta javana*	Diptera : Syrphidae	Tailed mealy bug
+*Brumoides suturalis*	Coleoptera: Coccinellidae	Tailed mealy bug
+*Genus nr. Scymnus*	Coleoptera: Coccinellidae	Tailed mealy bug
+*? Gitona* sp.	Diptera : Drosophididae	Tailed mealy bug
+*Leucopsis* sp.	Hymenoptera:Andrenidae	Tailed mealy bug
+*Mallada* sp.	Neuroptera: Chrysopidae	Tailed mealy bug
+*Scymnus* sp. A	Coleoptera: Coccinellidae	Tailed mealy bug
+*Scymnus* sp. B	Coleoptera: Coccinellidae	Tailed mealy bug
+*Spalgis epius*	Lepidoptera: Lycaenidae	Tailed mealy bug

+*Triommata*	Diptera: Coccinellidae	Tailed mealy bug
+*Coccidivora*	Coleoptera: Coccinellidae	Tailed mealy bug
* *Amyosoma leuzerae*	Hymenoptera: Braconidae	Red borer
* *Closterocerus insignis*	Hymenoptera: Eulophidae	Leaf miners
* *Halticoptera* sp.	Hymenoptera: Pteromalidae	Leaf miners
* *Herbertia indica*	Hymenoptera: Pteromalidae	Leaf miners
* *Opius* sp.	Hymenoptera: Braconidae	Leaf miners
* *Pediobius* sp.	Hymenoptera: Eulophidae	Leaf miners
* *Sphegigaster* sp.	Hymenoptera: Pteromalidae	Leaf miners
* *Teleopterus* sp.	Hymenoptera: Eulophidae	Leaf miners
* *Anicetus annulatus*	Hymenoptera: Eulophidae	Green scale
* *Aprostocerus purpureus*	Hymenoptera: Eulophidae	Green scale
* *Cerapteroceroides* sp.	Hymenoptera: Encyrtidae	Green scale
* *C. nr. similis*	Hymenoptera: Encyrtidae	Green scale
* *Cheiloneuromyia Javensis*	Hymenoptera: Chalcidae	Green scale
* *Coccophagus Bogoriensis*	Hymenoptera: Aphelinidae	Green scale
* *Coccophagus cowperi*	Hymenoptera: Aphelinidae	Green scale
* *Coccophagus lycimnia*	Hymenoptera: Aphelinidae	Green scale
* *Coccophagus* sp.	Hymenoptera: Aphelinidae	Green scale
* *Encyrtus lecaniorum*	Hymenoptera: Encyrtidae	Green scale
* *Marietta javensis*	Hymenoptera: Aphelinidae	Green scale
* *Microterys flavus*	Hymenoptera: Encyrtidae	Green scale
* *Prochiloneurus* nr.	Hymenoptera: Encyrtidae	Green scale
+*Chilocorus nigritus*	Coleoptera: Coccinellidae	Green scale
+*C. circumdatus*	Coleoptera: Coccinellidae	Green scale
+*Menochilus*	Coleoptera: Coccinellidae	Green scale
+*Sexmaculatus*	Coleoptera: Buprestidae	Green scale
+*Jauravia pallidula*	Coleoptera: Coccinellidae	Green scale
+*Jauravia pubescens*	Coleoptera: Coccinellidae	Green scale
+*Micraspis cardoni*	Coleoptera: Coccinellidae	Green scale
* *Anysis alcocki*	Tridymidae: Chalerdoidae	Brown scale
* *Coccophagus*	Hymenoptera: Aphelinidae	Brown scale
* *Anthrocinus*	Hemiptera: Anthocoridae	Brown scale
* *Coccophagus cowperi*	Hymenoptera: Aphelinidae	Brown scale
* *Encyrtus lecaniorum*	Hymenoptera: Chalerdoidae	Brown scale

+*Chrysopa sp.*	Neuroptera: Chrysopidae	Brown scale
+*? Dicrodiplosis fulva*	Diptera: Cecidomyiidae	Brown scale
+*Eublema* sp.	Lepidoptera: Noctidae	Brown scale
+*Megommata* sp.	Diptera: Cecidomyiidae	Brown scale
* *Agrypon* sp.	Hymenoptera: Ichneumonidae	Hairy caterpillars
* *Anastatus* sp. nr	Hymenoptera: Eupelmidae	Hairy caterpillars
* *Kashmirensis*	Ephemenoptera: oligoneriidae	Hairy caterpillars
* *Apanteles* sp.	Hymenoptera: Braconidae	Hairy caterpillars
* *Blepharisna* sp.	Protista: Heterotrichidae	Hairy caterpillars
* *Coccigomymus* sp.	Hymenoptera: Braconidae	Hairy caterpillars
* *Dicamptus* sp.	Hymenoptera: Ichneumonidae	Hairy caterpillars
* *Melonichneumon (Barichneumon)* sp.	Hymenoptera: Ichneumonidae	Hairy caterpillars
* *Pimpla* sp.	Hymenoptera: Ichneumonidae	Hairy caterpillars
* *Telenomus* sp.	Hymenoptera: Scelionidae	Hairy caterpillars
* *Villa* sp.	Diptera: Bombyliidae	Hairy caterpillars

+ Predator; * Parasitoid

A complex of parasitoids and predators naturally regulate the populations of scale insects and mealy bugs on coffee. The common predators of mealy bugs are *Triommata coccidivora, Cacoxenus perspicax, Spalgius epius, Nephus* sp. *Pseudoscymnus* spp. *Pullus pallidicomis* and *Horniolus* sp. The green scale, *Coccus viridis* has two main parasitoids *Coccophagus cowperi* and *C. bogoriensis*. The former is also a major natural biocontrol agent of *Saissetia coffeae* which causes up to 92% parasitization. The coffee borer, *Hypothenemus hampei* which got introduced in to India in 1990 has no indigenous natural enemies and therefore, three exotic parasitoids viz. *Cephalonomia stephanoderis, Prorops nasuta* and *Phymastichus coffea* were introduced of which only the last survived under the Indian conditions. This parasitoid has been able to establish in several areas.

The hairy caterpillars, *Eupterote canarica* and *E. fabia* are polyphagous and feed not only on coffee leaves but also on the leaves of shade trees and cardamom plants. Other hairy caterpillars seen on cardamom include *Eupterote cardamomi, E. testacea, Lenodera vitata* (Eupterotidae) *Euproctis ustifolia* (Lymantridae) and *Alphaea biguttata* (Arctidae). The hairy caterpillars can completely defoliate shade trees like *Bishofia javanica* and *Syzygium jambolana.*

White stem borer, though mostly confined to arabica coffee can also live in robusta plants, tree coffee and teak trees. However, it does not breed in these plants. The berry borer takes shelter in the seeds of several plants such as *Crotalaria, Lantana, Maesopsis,* tamarind and tea but will not feed or breed in these seeds. Coffee shot hole borer has several alternate host plants such as dapaps, *Annota,* Avocado, Cocoa, Crotalaria, Mango, Neem, *Dendrobium, Clerodendron* etc. The important larval ectoparasite is *Tetrastictus xyleborum*.

Cocoa crop was introduced in India in the early part of the 20th century, but commercial cultivation started in 1970. Cocoa is usually planted under coconut and arecanut plantations, where more light will penetrate through the canopy. The density of Cocoa plants will be slightly higher inside the areca gardens. The crop is grown almost exclusively in homestead gardens in Kerala where other crops such as coconut, arecanut and bananas are also planted in a mixed cropping system. Being an introduced crop, many indigenous pests have colonized this plant. Mealy bugs often colonise cocoa plants and among them *Planococcus lilacinus* and *P. citri* are more important *P. lilacinus* is a serious pest causing damage to cocoa throughout the country. The mealy bug is present throughout the year on the plant, invades all the plant parts such as the tender shoots, flower cushions, young cherells as well as mature pods. Mirid bugs form yet another group of serious pests. *Helopeltis theivora* affecting tea and *H. antonii* affecting cashew trees feed on cocoa plants. Similarly, coreid bugs are also well known cocoa pests. Among the homoptera adults and nymphs of the membracid, *Gargara* sp. infest young cherelles and shoots. The aphid *Toxoptera aurantii* occurs on flowers, flowers buds, flush leaves and tender stems. Again, *T. aurantii* occurs on tea shoots and as in tea they are also attacked by syrphid, coccinellid and hemerobiid predators on cocoa plants. The red banded thrips, *Selenothrips rubrocinctus* is very common throughout the tropics on cocoa and on several other plants.

Cocoa trees exhibit periodic flushing which is attractive to foliage feeding insects, especially caterpillars and beetles. However, the caterpillars restrict their damage the beans and do not cause serious harm to cocoa plants. However, certain psychids can cause defoliation. The hairy caterpillars (Lymantridae) affecting cocoa are many and these species are polyphagous and cause damage to coffee and cardamom plants. The more important genera are *Lymantria, Euproctis* and *Dasychira. Lymantria ampla* is common on Cocoa. The caterpillars, seen abundantly after rains cause severe damage to leaves of young plants. The early instar caterpillars feed on leaves or the surface tissues of growing pods, during day and night, but later instar caterpillars are nocturnal in habit. During daytime they hide on fallen leaves at the base of the tree or at the base main stem. The indigenous natural enemies of this lymantriid include a pupal parasite and a braconid larval parasite *Apanteles* sp. Caterpillars of *Euproctis* spp. are also defoliators of cocoa. They feed on tender leaves, surface of cherelles and green pods. The sporadic incidence of high population of *Euproctis* spp. can cause severe damage of tender leaves.

Grubs of many scarabaeid beetles feed on root and decaying organic matter in the soil. The adult beetles feed on soft leaf tissues and flowers of cocoa. Leaf eating ash weevils, mainly *Myllocerus* spp. feed on tender leaves making irregular holes. They make sporadic appearance in some areas after rains and cause serious damage. Grubs of these beetles frequent the soil. *M.viridanus* Fab. and *M. maculosus Desb.* (Curculionidae) are the two important species of which *M. viridanus* occur as a major pest in all cocoa growing tracts of Kerala and Tamil Nadu. Adults are seen in large numbers on the lower surface of leaves and feed on the interveinal tissues leaving the veins intact. Population peak is observed in July-September. Infestation is relatively severe on young plants and quite often, the entire foliage of such plants are skeletonised leading to growth retardation. The incidence of ash weevils is more serious in coconut-cocoa plantations.

Like the leaf feeding beetles and weevils, stem boring caterpillars and beetles are seen more in forest cleared areas or plantations adjacent to forests in the cocoa growing regions. Coleopteran and lepidopteran borers attack cocoa, especially the seedlings and trees planted near forests. The red borer of coffee, *Zeuzera coffeae Nietner* (Cossidae) is a pest and this species bore in to the stems of young tea and coffee plants also.

The system of mono cropping in beverage crops results in the persistence of the most favoured insect which in turn is aggravated by inadequate control measures and destruction of the environment, introduction of susceptible cultivars and unscientific cultivation practices.

Acknowledgement

Information relating to cocoa crop has been mostly obtained from the website of CPCRI http://cpcri.nic.in/cropprotection1pest.htm

References

Anonymous 2003. *Coffee Guide* pp 200 Central Coffee Research Institute, Chikmagalur, Karnataka, India.

Banerjee, B 1983. Arthropod accumulations on tea in young and old habitats *Ecol. Entomol.*

Banerjee, B, 1981. An analysis of the effect of latitude age and area on the number of arthropod species of tea *J. appl. ecol* 18:339-342.

Barua, D.N. 1989, *Science and Practice in Tea Culture.* Tea Research Association, Calcutta, pp.509.

Chen, Z.M. and Chen X.F. 1989a. An analysis of world tea pest fauna. *J. Tea Sci.,* **9** (1) 73-88.

Gurusubramanian, G, Senthilkumar, N, Tamuli, AK, Sarma, M, Rahman, A, Bora, S, and Roy, S. 2008 Biointensive integrated management of the tea pests for sustainable tea production in north east India *Int. J. Tea Sci* **7** (3&4):45-59.

Muraleedharan, N and Chen Z.M 1997 Pests and disease of tea and their management *J. Plantation Crops.* 25(1):15-43.

Muraleedharan, N and Selvasundaran, R. 1991 Bioecology of *Phytodietus spinipes* (Cameron) a parasitoid of *Homona coffearia* Nietner the tea tortrix *J. Plantn. Crops* 19(1):26-32.

Muraleedharan, N, Selvasundaram, R and Radhakrishnan, B 1988 Natural enemies of certain tea pests occurring in southern India *Insect Sci. Appl.* 9 (5):647-654.

Muraleedharan, N, Selvasundaram, R and Radhakrishnan, B, 2001. Parasitoids and predators of tea pests in India. *J. Plantation Crops* 29(2):1-10.

Nagarkatti, S, Sankaran, T and Majunath, T.M. 1979, Natural control of the scale *Fiorinia theae* Green Coccoidea: Diaspididae) in Assam *Protn. Ecol.* **1**:179-184

Radhakrishnan, B. Loganathan, S. and Muraleedharan, N. 1988 Coccinellid predators of *Toxoptera aurantii* (Boyer de Fonscolombe) infesting tea in southern India *Newsl. appl. Soc. India* 7(1):7-8.

Radhakrishnan, B and Muraleedharan, N. 1993. Bioecology of six species of syrphid predators of the tea aphid *Toxoptera aurantii* (Boyer de Fonscolombe) in southern India *Entomon* 18(2/3):175-180.

Rahman, A, Sarma, M, Borthakur, M, Gurusubramanian, G, 2005. Integrated management of tea pests-1 *Helopeltis,* thrips, green fly, looper, red slug bunch caterpillar and nuisance pests pp. 27-32 In: *Plant Protection in Tea* Eds. Dutta, AK, Gurusubramanian, G and Borthakur, B.K. Toklai Experimental Station, Assam Printing Works Pvt. Ltd., Jorhat, Assam.

Ramaiah, P.K. and Raman, K. 1967 Studies on the dieback in coffee (*Coffea Arabica L.*) III. Critical evaluation of the factors causing the new malady *Indian Coffee* **31** (3): 7-11.

Roy, S, Talukdar, T, Saha, AB, Banerjee, DK, Sannigrahi, S and Gurusubramanian, G 2005 Species richness and seasonal abundance of spiders and lad bird beetle fauna in the tea ecosystem of north Bengal pp. 347-351. In: *Proceedings of the 34th Tocklai Conference*, Tea Research Association, Jorhat, Assam.

Somasekhar, P. 1958 Pests of coffee and their control. *Indian Coffee* 22(6):220-248.

Chapter - 5

Biodiversity of Insect Biocontrol Agents

R.J. Rabindra, C.R. Ballal, S.K. Jalali and T. Venkatesan
National Bureau of Agricultural Important Insects, ICAR, Bangalore (Kar.)

The total number of existing insect species is estimated at 1.84 to 50 million, while that of described insect species in the world is estimated at 7,50,000 to over 1 million. Due to the existence of many sibling species, the accuracy of these estimates is doubtful. Against a background of millions of years of lesser climatic adversity in the tropics, many plant species and complex plant architectures have evolved. These factors coupled with plant-insect co-evolutionary changes, have contributed to the vast insect variety in the tropics. India has about 10% of the world fauna of insects and is one of the megadiversity nations in the world in terms of insect diversity. Our insect biodiversity is unique as evidenced by the high level of endemism. Nearly one-third of our insect fauna known so far is endemic (Varshney, 1998). For research and development of sustainable pest management systems to combat pest-induced losses in tropical agriculture, sound knowledge of biosystematics of natural enemies is of fundamental importance so that the potential ones could be identified either for conservation strategies or for utilization in biological control. The history of biological control abounds with numerous examples of failed programmes, which could probably be due to inaccurate identifications of pests and / or their natural enemies.

There are three main levels of diversity: genus, species and ecosystems (Reid and Miller, 1989; Solbrig, 1991) and diversity in each level is important in an agro-ecosystem. In a functional agro-biodiversity programme, the aim is to increase the number of beneficial organisms and decrease the number of detrimental ones. Parasitoids and predators are known to restrict the population growth of the pest/host. However, they act differently depending on the population level of the host. Parasitoids and predators can be conserved, preserved and multiplied under laboratory conditions or in commercial production units for field release against target pests or diseases. The role played by natural enemies in nature becomes evident when insecticide use is stopped or reduced. By reducing the use of chemical pesticides, the indigenous bio-diversity is maintained and extra control is achieved by preserving the natural enemies.

Natural biodiversity agroecosystems

It is well known that a high degree of inter and intra species diversity of entomophagous insects exist in different crop ecosystems. In corps like rice, cotton and certain vegetable crops, one can find as many parasitoids and predators as the number of pests, often resulting in a steady state equilibrium. Several factors both biotic and abiotic, may alter this sensitive balance. Changes in the crop diversity, sequence, host plant resistance, tritrophic interaction and hyperparasitoids may reduce the diversity of entomophagous insects. Abiotic factors that affect the diversity include climate change, erratic monsoon, extreme temperature, cyclonic weather and use of broad-spectrum chemical pesticides.

In general, high and prolonged levels of diversity in an agro-ecosystem will permit the development of a higher level of internal links, which will promote stability. However, the nature and quality of the trophic interactions and responses could be the key factors for stability in agro-ecosystems.

There are several potential parasitoids in nature which are important mortality factors of major pests in India. Over a dozen parasitoids have been recorded on the citrus mealybug *Nipaecoccus viridis*. However, *Anagyrus dactylopii* was the dominant one, parasitising up to 90 per cent in the field (Ali, 1957; Subba Rao *et al.,* 1965). On citrus butterfly *Papilio demoleus* Linnaeus, egg parasitoid *Trichogramma chilonis* Ishii parasitised up to 76 per cent and *Telenomus* sp. nr. *incommodus* up to 78 per cent in February (Krishnamoorthy and Singh, 1988; Jalali and Singh, 1990). *Distatrix papilionis* is a dominant parasitoid of *P. demoleus* and *T. chilonis, T. incommodus* and *D. papilionis* caused a cumulative parasitism of 88 per cent (Krishnamoorthy and Singh, 1988). *T. chilonis, Melalophacharops* sp. and *D. papilionis* could be utilised for the biological suppression of citrus butterfly. The eggs of fruit sucking moth, *Othreis fullonia* are successfully parasitised by *T. chilonis*, which suggests the possibility of utilising *T. chilonis* for the control of this pest (Dodia *et al.,* 1986).

On cabbage, cauliflower and other cole crops, diamondback moth (DBM), *Plutella xylostella* is a major pest. At Anand, Yadav *et al.* (1975) recorded up to 72% parasitism by *Cotesia plutellae.* In Karnataka and Tamil Nadu, *C. plutellae* was known to cause up to 80% parasitism (Jayarathnam, 1977; Nagarkatti and Jayanth, 1982). In the Nilgiris, *Diadegma semiclausum* could parasitise 2.32 to 68% of *P. xylostella* larvae (Chandramohan, 1994).

In several ecosystems, the naturally occurring bio-agents are able to manage the pests, provided their parasitising/ predatory ability is not hampered through chemical pesticide interventions. Two striking examples have been observed in the sugarcane and paddy ecosystems. The sugarcane woolly aphid, *Ceratovacuna lanigera,* was observed as a serious pest of sugarcane and reported in outbreak proportions from western and southern India (Rabindra *et al.,* 2002; Joshi and Viraktamath, 2004). The parasitoids which were recorded on this pest in Nagaland included *Aphelinus desantisi, Encarsia falvoscutellum, Diaeretiella rapae, Anagyrus* sp. and *Antocephalus* sp. (Tripathi, 1995). In Jorhat in Assam, *Encarsia flavoscutellum* was observed in abundance parasitising the sugarcane woolly aphids. The heavy incidence of this parasitoid could manage the woolly aphid population. This parasitoid was re-distributed in several states where the pest was a major problem. The parasitoid could establish well and along with the naturally occurring populations of the predators *Dipha aphidivora* and *Micromus igorotus* could bring the pest under control, provided application of chemical pesticides was withdrawn.

The natural biodiversity was conserved in 1250 hectares of paddy in kole lands of Adat Panchayat in Thrissur District in Kerala by avoiding the application of chemical pesticides and adoption of bio-intensive pest and disease management. This led to the enhancement of bio-diversity in rice farms, evident through an increase in the population of birds, predatory insects, fishes, etc. and a reduction in insect pest and disease incidence. These are striking examples to highlight the benefits of conservation of the bio-diversity of natural enemies in different agro-ecosystems.

Impact of host plants

Host plants exert a major influence on the performance of several parasitoids and predators. The activity of egg parasitoids is almost nil in pigeonpea crop and totally absent in chickpea due to the presence of glandular and non-glandular trichomes mainly on the surface of the calyces and pods and due to the presence of acidic exudates. Several plant characters are known to affect the searching behaviour and parasitization efficiency of *Trichogramma* spp. (Romeis *et al.,* 1998, 1999a & b; Ballal and Singh, 2003). The potential indigenous larval parasitoids that have been recorded on *H. armigera* in the pigeon pea and chickpea ecosystems are the ichneumonid early larval parasitoids *Campoletis chlorideae, Eriborus argenteopilosus* and tachinid late larval/larval-pupal parasitoids, *Goniophtalmus halli, Senometopia (Carcelia) illota* and *Palexorista laxa* (Bilapate *et al.,* 1988). These have proved to be potential mortality factors of the pest. The lab-reared parasitoid *Campoletes chlorideae* which were originally from the pigeonpea ecosystem could not efficiently parasitise *Helicoverpa armigera* larvae from the cotton ecosystem, whereas the parasitoids from the cotton ecosystem were capable of parasitising more than 40% of the larvae of cotton ecosystem (Ballal *et al.,* 2001). This clearly indicated that the performance of *C. chlorideae* was largely governed by the host plants on which the pest was found. Bajpai *et al.* (2002) reported that on chickpea plants, the chemical cues released during feeding by the *H. armigera* was essential for *C. chlorideae* to be attracted to the infested plants and to induce parasitism. Parasitism was also governed by host plant variety (Ballal and Gupta, 2003).

Strain variations in natural enemies

Genetic diversity within a species can be of considerable importance to agro-biodiversity projects. Several examples have been quoted by Unruh and Messing (1993). The intraspecific variations in several species of natural enemies became evident only when they were attempted to be used in biocontrol projects.

Populations of *Trioxys pallidus* (a parasitoid of the walnut aphid *Chromaphis juglandicola*) from the cooler Mediterranean region (France) established well in the cooler southern coastal regions of California while populations obtained from Iran which had evolved to more extreme climatic conditions could establish in the harsher interior regions of California. The larch sawfly *Pristophora erichsonii* invaded North America from Europe and became a major forest pest. An ichneumonid *Mesoleius tenthredinis* was introduced from Europe and provided effective control of the pest. This control was negated when a resistant strain of the larch sawfly appeared and spread. This strain could encapsulate and kill the parasitoid egg. This defense mechanism was overcome by the importation and colonization of a Bavarian strain of *M. tenthredinis,* which could overcome the encapsulation reaction (Turnock *et al.,* 1976).

Studies conducted in India have indicated that intra species variations occur in several parasitoids and predators, either based on their geographic location or the host plant on which they are found or the host/pest which they attack. Genetic resources may also be accessed from beyond political boundaries. Three strains of *Encarsia perniciosi viz.,* Californian, Russian and Chinese, were introduced for the control of *Quadraspidiotus perniciosus.* In addition, *A. diaspidis* (origin: Japan) was introduced from California. All the strains could establish and the Russian strain of the parasitoid gave 89 per cent parasitism in Himachal Pradesh. *A. diaspidis* in combination with *E. perniciosi* provided 86.5 per cent parasitism. In Kashmir, the Russian and Chinese strains appeared to be superior. American and Chinese strains of *E. perniciosi* were also released in the Kumaon hills of Uttar Pradesh and the population of the pest was reduced by about 95 per cent.

Strains of *Campoletis chlorideae* obtained from different parts of India were evaluated for their biological parameters. It was found that the Sehore strain was most efficient (Ballal and Ramani, 1994). Variations were observed in the performance of *C. chlorideae* populations collected from different crop ecosystems. Intra species variations have been recorded in *Chrsoperla carnea* group, *Cryptolaemus montrouzieri, Goniosus nephantidis, Trichgramma* spp. and *Cotesia flavipes.* Populations of these natural enemies varied in their biological attributes as well as tolerance to high temperature and chemical pesticides through a natural selection process.

Development of superior strains of natural enemies and their exploitation

Studies have indicated that superior strains of natural enemies could be developed, which would be better performers in the field situations. *Trichogramma chilonis* for example is widely used as an effective biocontrol agent against different lepidopteran borers in various ecosystems namely cotton, rice, sugarcane, vegetables and fruits. The trichogrammatids are susceptible to a broad spectrum of insecticides. With the increasing preference for sustainable agriculture and awareness of side effects of indiscriminate use of insecticides, biological control agents are attracting more attention. However, use of chemical pesticides continues to be widely adopted for insect pest control. In such a crop scenario where insecticides are frequently used, releases of susceptible strain of *T. chilonis* will not be able to provide any appreciable control. Hence, attempts were made to develop insecticide tolerant strains of parasitoids.

Endosulfan tolerant strain of *T. chilonis* (ENDOGRAM)

An endosulfan-tolerant strain of *T. chilonis* was developed at the National Bureau of Agriculturally Important Insects (formerly Project Directorate of Biological Control). The improved strain is distinctly superior to the normal strain of *Trichogramma*. The strain was sold to M/S Excel industries Limited, Mumbai for commercialization and the strain has been named as ENDOGRAM (Jalali *et al.*, 2006). The performance of the endosulfan tolerant strain of *T. chilonis* was evaluated against *Helicoverpa armigera* eggs on cotton plants and it was observed that under pesticide pressure, the tolerant strain could provide significantly higher parasitism in comparison to the susceptible strain (Jalali *et al.*, 2006; Ballal *et al.*, 2009).

Multiple insecticide tolerant strain of *T. chilonis*

This strain was further improved for tolerance to multiple pesticides (endosulfan, monocrotophos and fenvalerate) (Provisional Patent No. 850/Del/2003: Dated 27/06/2003). Trials conducted utilising the multiple insecticide- tolerant strain (MITS) of *T. chilonis* against *H. armigera* in Gujarat, Karnataka and Tamil Nadu on cotton revealed that the MITS could provide 63 % more parasitization of *H. armigera* compared to the susceptible laboratory strain. This invention has great potential to be utilized commercially on crops like cotton, tomato, cabbage, rice, etc. where the above insecticides are used on a very large scale. This could also lead to reduction in overall load of insecticides on these crops, thereby protecting the crop, reducing deleterious environment effects and producing insecticide free crop to sustain the action of natural enemy biodiversity.

Pesticide tolerance in green lace wing, *Chrysoperla* spp. (*carnea*-group)

Chrysoperla spp. (*carnea*-group) (Chrysopidae: Neuroptera) is one of the important biological control agents and used effectively to manage various insect pests especially sucking pests in different agro-ecosystems (Canard *et al.,* 1984; Venkatesan *et al.,* 2000; Athan *et al.,* 2004). Chrysopid predators are also available in commercial insectaries for augmentative releases in biological control programmes in different countries. *Chrysoperla* occurring in most of the regions in India were identified as *Chrysoperla zastrowi arabica* (Venkatesan *et al.,* 2008 and Venkatesan *et al.,* 2009). These predators known to be relatively resistant to some chlorinated hydrocarbons, pyrethroids and microbial insecticides, however, are susceptible to most of the commonly used organophosphorus insecticides (Grafton-Cardwell and Hoy, 1985). Twenty-four hour Bioassay technique IOBC/WPRS Working Group) revealed that there was variation among different populations for tolerance to monocrotophos. The population collected from Sriganganagar had highest resistance (1819.72 ppm), followed by Delhi (1564.05 ppm). Populations collected from lab reared and Shimla was highly susceptible to monocrotophos with LC_{50} of 4.02 ppm and 5.86 ppm, respectively. Of the 9 field populations, Sriganganagar population exhibited 452.66 folds resistance compared to laboratory population, however, Shimla was found to be most susceptible (1.46 fold resistance (Venkatesan *et al.,* 2009). Besides, populations of *Chrsoperla* collected from different locations in India varied in their esterase activity indicating genetic diversity which can be exploited for development of pesticide tolerant strains.

Pesticide tolerance in *Cryptolaemus montrouzieri*

Populations of *C. montrouzieri* from Shimoga, New Delhi and Coimbatore varied in their tolerance to imidachloprid, deltamethrin, cypermethrin and quinalphos. All the populations were able to withstand imidacloprid. Quinalphos caused 100% mortality of lab reared, Coimbatore, Delhi and Shimoga populations. Bangalore population was relatively tolerant. Shimoga population was relatively tolerant to deltamethrin followed by Coimbatore.

Temperature tolerance in *Chrysoperla* spp. (*carnea*-group)

Studies on high temperature tolerance (32-38°C) in different populations of *Chrysoperla* revealed that at the above high temperatures, adult emergence of the populations from Delhi and Punjab was 75.0 and 74.0 days, respectively and the same of the lab reared was only 60.0 days. Fecundity of the predator from different locations varied from 38 to 121.0 and maximum was in the case of Punjab followed by Delhi and lab reared populations. Longevity of these populations varied from 34.0 to 36.0 days. Field collected population survived better at variable high temperature compared to lab reared population.

Genetic variation and molecular characterization of natural enemies

Markers are powerful molecular tools that have facilitated the study of diverse areas of biology, both pure and applied, covering questions of phylogeny, evolution, ecology and population dynamics. The field of molecular biology has expanded greatly in the last ten years and many entomologists now wish to use this technology, since it is a new level of resolution for the study of insect ecological systems and taxonomy. It is important to understand the basic processes in molecular biology and their potential application, i.e. the sort of study each marker is best suited for. Successful biocontrol depends on the selection of the most suitable species or strain of natural enemy. The most successful natural enemies are highly host-specific. Therefore, correct identification is extremely important, when either the pest or the natural enemy has closely related species that are morphologically indistinguishable (cryptic species). Identification of distinct biological and biochemical differences among populations of natural enemies has important consequences for future mass production and field release.

Some strains are better suited to standard rearing on host insects than others. Furthermore, some strains can change more rapidly through continued rearing and selection in the laboratory than others. Therefore, recognition of species and strains with desirable traits will have better impact on the successful multiplication of the parasitoid and success of the parasitoid or predator as a biocontrol agent. The identification of different species and strains may also enable us to narrow the host range of parasitoid/predator species and to direct it more specifically against a particular species of pest. Molecular techniques could solve, in part, the problem of inadequate systematics for some natural enemy groups. The taxonomic level at which specific genes or nucleotide regions are useful varies across taxa. Different genes have been used in order to evaluate the genetic relatedness of closely related species or populations. In theory, rapidly evolving genes and nucleotide regions are useful for comparisons of closely related taxa. Molecular markers are commonly used by biologists to perform a number of tasks, including the genetic fingerprinting of insect strains or species, plant varieties, determining similarities among inbred varieties, mapping of plant genomes, and establishing phylogeny among plant species. Application of molecular markers in biological control is still in its infantile stage.

Molecular markers are a string of nucleic acids, which makes up a segment of DNA and are near the DNA sequence of the desired gene. Molecular markers and the genes are close together on the same chromosome, hence tend to remain close together generation after generation. There are two types of molecular markers being used i.e. Isozyme markers and DNA based markers

Isozyme markers

Proteins are final products of a gene. Changes in coding sequence will result in corresponding changes in primary structure of the protein. Even a single change in amino acid can have a marked effect on the migration of the protein under an electric field. These isozymes differ in their charge and hence mobility. In Isozyme analysis crude sample extracts are subjected to electrophoresis on starch or on polyacrylamide gels. Following electrophoresis, the enzymes of interest are detected by treating the gels with specific activity stains. Variation in banding patterns obtained between individual samples can be used to sort out genetically the samples tested. The advantage is that the technique is relatively simple and less expensive. The disadvantages include availability of limited number of enzyme loci and expression pattern is influenced by physiological and environmental factors (Loxdale, 1994).

DNA-based molecular markers

DNA-based molecular markers have acted as versatile tools in various fields like taxonomy, physiology, embryology, genetic engineering, etc. They are no longer looked upon as simple DNA fingerprinting markers in variability studies or as mere forensic tools. Ever since their development, they are constantly being modified to enhance their utility and to bring about automation in the process of genome analysis. A number of DNA based marker systems like AFLP, RAPD, RFLP, SCAR, SSCP, SSR and a few others like the micro satellite markers are now available.

Species possessing minor morphological differences can be identified easily using a combination of PCR reaction in which the ITS2 spacer is amplified followed by determination of the size of the spacer by gel electrophoresis and in some cases restriction digestion of the PCR product. Two types of rapidly evolving genes and nucleotide regions used for comparison of closely related species are mitochondrial cytochrome oxidase gene and internal transcribed spacer gene.

Mitochondrial cytochrome oxidase gene

Mitochondrial genes have been used extensively in evolutionary studies because of their high rate of evolution, maternal transmission, no recombination and the belief that most of the variations within species are selectively neutral.

Characterization of Internal transcribed spacer (ITS), RFLP and RAPD in natural enemies

Trichogramma

Taxonomy of minute parasitoid wasps of the genus *Trichogramma* traditionally has been difficult. Their small size and lack of distinct morphological characters made identification a difficult task. Nagarkatti and Nagaraja (1977) showed the value of male genitalia for species characterization. Since then many species have been described and specialists are now able to identify most species using morphological characters. However, morphological identification requires slide preparation which is time consuming and is a difficult task (Platner *et al.,* 1999). Morphological identification also depends on the availability of males and requires considerable expertise for identifying minor differences in order to separate species. Thus, a great need has been felt for identifying a simpler and more applicable identification system. Stouthamer *et al.* (1998) have shown that the DNA sequence of the internally transcribed spacer (ITS-2) of *Trichogramma* wasps can be used for species identification.

In eukaryotes, the genes encoding18S and 28s RNA are clustered in tandem repeats in nuclear genome (Hoy, 1994). The ITS region between 18Sand 28Scoding region usually have higher degree of polymorphism than the coding region (Hoy, 1994), hence ITS region has been extensively used in examination of taxanomic status of species and for diagnostic purpose

ITS-2-RFLP method was employed to distinguish 12 different trichogrammatids comprising indigenous and exotic species such as *Trichogrammatoidea armigera* and *Tr. bactrae, Trichogramma achaeae, T. chilonis, T. japonicum, T. embryophagum, T. pretiosum* (Thelytokous Form-TF), *T. brassicae, T. dendrolimi, T. evanescens* and *T. mwanzai*. ITS-2 region was amplified and complete ITS-2 sequences of the nine species were deposited in Genbank. The size of the amplified product ranged from 500 to 900 bp. Restriction enzyme digestion of ITS-2 region showed different banding profiles for these 12 species. Dichotomous keys using the size of the ITS2 product and the restriction fragment length polymorphism for the enzymes (ECOR1, Mse1, Mva1 and Taq1) allowed quick species identification of these trichogrammatids (Ashok *et al.*, 2009)

Chrysoperla

Molecular characterization of ITS-2 region of *C. carnea* (538bp) was done and the DNA sequence was submitted to NCBI, USA. GenBank Accession No. DQ825504 (Venkatesan *et al.*, 2008). RAPD assay involved genomic DNA of *C. carnea* collected from Punjab and Delhi using short primers CO7, C12, C14, C15, C16, C18, C19, KO1, KO2, K15 and K19 (9-10 bases). Amplified products were separated on 1.2% agarose gels and their detection by ethidium bromide staining was done. Random primers C16, C19, KO2 and K19 produced polymorphic banding patterns, which can be used to distinguish the two populations of *C. carnea* collected from Punjab and Delhi.

Cryptolaemus montrouzieri and *Goniozus nephantidis*

Isolation and purification of DNA of *C. montrouzieri* was done by CTAB method. ITS-2 regions of *C. montrouzieri* collected from Coimbatore, Delhi and Shimoga were amplified and there was no significant difference in base pairs among these populations. Molecular characterization of ITS-2 region (800–900 bp) of *G. nephantidis* and RAPD-PCR assay of the different populations revealed genetic variability among them. The similarity coefficient matrix indicated Kerala and Karnataka populations to be 86% similar, while those from Andhra Pradesh, Gujarat and Tamilnadu were distinct with similarity in the range of only 10-13%.

Wolbachia-induced genetic diversity

Endosymbionts are associated with thelytoky in several species of hymenoptera while not all thelytokous species have a microbial cause. Bacteria of the genus Wolbachia was found in thelytokous *Trichogramma* spp (Stouthamer & Werren, 1993). In addition, circumstantial evidence suggests that infection with the Wolbachia is widespread in other thelytokous species such as *Encarsia formosa.* Wolbachia can cause thelytoky, male killing, cytoplasmic incompatibility, and female mortality in arthropods (Stouthamer 1997, Vavre *et al.,* 2000). For mass rearing, thelytoky has an advantage over arrhenotoky in that all the hosts would be used for the production of female wasps. Females being the effective biocontrol agents, the use of thelytokous wasps can result in a reduction in cost of production and field utility. Further the higher rate of population increase in thelytokous forms is an added advantage. Thus it is speculated that using the thelytokous forms of a wasp species may be more effective for biological control than using the arrhenotokous form. At NBAII, studies on endosymbionts in Trichogrammatids revealed that wolbachia is present in *T. corbudensis* (France), *T. semblidis* (France); *T. evanescens* (France); *T. pretiosum* (USA) and *T. pretiosum* (France) using Wsp primer (bp- 550-600 bp) and FtsZ primer (700-750 bp). Studies on horizontal transmission of wolbachia in non-wolbachia Trichogrammatid species (*T. acheae* and *T. japonicum*) increased female proportion after F15 generations (PDBC, 2009).

Conservation Biological Control

Conservation of the diversity of natural enemies is probably the most important, readily available, generally simple and cost-effective strategy for pest control. Natural enemies occur in all production systems, from the backyard garden to the commercial field. They are adapted to the local environment and to the target pest. With relatively very little effort, the activity of these natural enemies can be enhanced. For example, parasitized aphid mummies are almost always present in aphid colonies. These natural controls are important and need to be conserved and considered when making pest management decisions. In many instances, the importance of natural enemies has not been adequately studied or does not become apparent until insecticide use is stopped or reduced. Often, the best we can do is to recognize that these factors are present and try to minimize the negative impacts on them.

Conservation biological control practices such as maintenance of refuges for natural bio-agents, conserving weed plants harbouring predators and egg parasitoids, use of safer pesticides, judicious and selective use of non-persistent pesticides, strip treatment, spot treatment, etc. were found to be effective conservation techniques in several crop ecosystems (Singh, 2002). Use of kairomones, synomones, pheromones, adjuvants, etc. to increase the searching ability and retention of parasitoids, building up of population of biocontrol agents by providing artificial structures, food, alternate host, attractive host plants, suppression of ants, etc., are some conservation techniques. Hyperparasitoids represent an addition to diversity, which has traditionally been considered as an adverse factor in agro-ecosystems. However, there is some indication that the addition of a hyper-parasitoid might increase stability and this might compensate for any adverse effect they may have in reducing the regulatory activity of primary parasitoids (La Salle, 1993).

However, the limiting factors for the abundance of parasitoids or predators are immensely varied or complex, even for one species in one area at one time. Hence it is a very difficult task to find the requirements for the survival of a single species. It may not be holistic for the realistic conservation of several insect species, which occur together in a single ecosystem. Landscape conservation and appropriate manage-ment is a more fruitful approach in such a situation.

Habitat manipulation is a useful approach, wherein the cropping system may be modified to favour or augment the natural enemies. Habitat manipulation techniques (to improve the population and performance of natural enemies) are easily incorporated into home gardens and even small-scale commercial plantings, but are more difficult to accommodate in large-scale crop production. There may also be some conflict with respect to pest control because of the difficulty in targeting the pest species as the refuges may be used by the pest insects as well as natural enemies. Habitat manipulation involves altering the cropping system to augment or enhance the effectiveness of a natural enemy. Many adult parasitoids benefit from sources of nectar and the protection provided by refuges such as hedgerows, cover crops and weedy borders. Mixed plantings and the provision of flowering borders can increase the diversity of habitats and provide shelter and alternative food sources.

Natural enemies can also be conserved by using insecticides or formulations which are least harmful and by timing applications to reduce the impact on beneficial arthropods. Natural enemy populations may be enhanced by increasing the diversity of plant species in the vicinity of the crop, changing cultural practices to ensure continuous availability of hosts and by providing alternative food sources (Pawar, 1986). Ballal and Singh (2001) reported non-intervention and thus conservation of natural enemies to be the best strategy for *Helicoverpa armigera* management in the sunflower ecosystem.

In terms of conservation biology, it has been clearly indicated that great abundance of an insect at one time does not preclude the possibility of its extinction in the future. The Rocky Mountain grasshopper, *Melanoplus spretus* is an example. This insect was a pest in the nineteenth century in the western states of North America and was extinct by 1902 (Lockwood and De Brey, 1990). Such instances could be occurring in the case of parasitoids and predators in India, but would have gone unnoticed. It is observed in several ecosystems that the abundance of some of the potential parasitoids and predators has drastically reduced, probably due to unfavourable human interventions. One example is the drastic reduction in the population of the naturally occurring potential larval ichneumonid parasitoid *Campoletis chlorideae* due to chemical pesticide interventions in the pulses ecosystem in our country.

Optimization of Insect Biodiversity in Agroecosystems

There are several methods by which agro-ecosystems can be manipulated to prevent pest build-up and enhance bio-diversity or efficacy of beneficial insects. A model project suggested by La Salle (1999) for optimization of insect bio-diversity in agro-ecosystems consists of the following steps:

1. Gather all pertinent published information for your system or organisms.
2. Involvement of taxonomists for authoritative identifications, before accessing literature for literature searches and for ascertaining alternative names of species under which information may be stored; seek help from

taxonomists for obtaining information on distribution and native homes of organisms, authoritative evaluation of information available in literature, methods for recognizing and distinguishing organisms and providing trainings and arranging workshops.

3. Building a knowledge base of the natural enemy complex for a given pest through information available in literature, specimen-based information available in museums and through field surveys. Information should include species of natural enemies, location, timing of their presence and estimated impact
4. Focus on sustainable use of preserved bio-diversity. Surveys to be conducted of the same and/or related pests and their natural enemies in native ecosystems so that the potential ones could be imported and utilized for biocontrol programmes.
5. The identification and utilization of resistant or tolerant cultivars of plants.
6. Identification of suitable crop management practices which are beneficial to natural enemies and detrimental to pests: eg. Crop species rotation, mixed and inter-cropping systems, selection of perennial rather than annual crops, selection of short maturing crops, synchr-onous and discontinuous planting, planting in seasons unfavourable to the pest, high planting density and small scattered fields, planning of time of first harvest, use of resistant cultivars, etc.
7. Environmental management and modification like providing artificial shelters, supplementary food and alternative hosts for natural enemies, improvement of pest-natural enemy synchronization, modification of adverse agricultural practices, maintenance of suitable adjacent habitats, etc.
8. Encouragement of growth of those weeds which provide resources for natural enemies
9. Review of the available methods and selection of the ones with potential value.
10. Documentation of diversity levels in disturbed *vs.* natural habitats, monocultures *vs.* polycultures and documentation of various conditions of within field habitat

modification. This would provide solid scientific basis for management decisions.

11. Information management and dissemination through publications, databases, websites and electronic publications.

Today, when advanced research is being conducted on developing natural enemies with superior traits for improved field performance, the effect of field releases of such strains on the existing bio-diversity would be an important area of research.

Promoting and protecting agro-biodiversity is an activity that requires widespread participation and support from farmers, researchers and environment protection groups. Government may not succeed in managing this task alone, even with the best regulations and controls. Input regulations can thus be designed to allow farmers and other stake-holders to take part so that interested and committed people outside the government are empowered to maximize their contributions towards promoting and protecting agro-biodiversity.

References

Ali, M.S. 1957. Some biological studies on *Pseudococcus vastator* (Maskell). *Indian Journal of Entomology*, **19**: 54-58.

Ashok Kumar, G., Jalali, S.K., Venkatesan, T., Stouthamer, R., Niranjana, P. and Lalitha, Y. 2009. Internal transcribed spacer-2 restriction fragment length polymorphism (ITS-2-RFLP) tool to differentiate some exotic and indigenous trichogrammatids egg parasitoids in India. *Biological control*,

Athan, R., B. Kaydan and M.S. Ozgokee, 2004. Feeding activity and life history characteristics of the generalist predator, *Chrysoperla carnea* (Neuroptera: Chrysopidae) at different prey densities. *Journal of Pesticide Scence,* 77: 17-21.

Bajpai, N.K., Ballal, C.R., Singh, S.P. and Srinivasan, R. 2002. Net house studies on performance of *Campoletis chlorideae* on *Helicoverpa armigera* larvae infesting chickpea. *Indian Journal of Applied Entomology* **16** (1): 22-24.

Ballal, C.R. and Gupta, T. 2003 Influence of chickpea varieties on ovipos-ition by *Helicoverpa armigera* and parasitising efficiency of *Campoletis chlorideae* p. 81, Presented at the *National Symposium on Bio-Management of Insect Pests* 29th to 31st March, 2003, Annamalai University, Annamalainagar, Tamil Nadu. 121 pp.

Ballal, C. R. and Ramani, S. 1994. Studies on four geographical strains of *Campoletis chlorideae* Uchida (Hymenoptera:Ichneumonidae). Presented at the *Fifth National Symposium on Advances in Biological Control of Insect pests* held by U.P. Zoological Society at Muzaffarnagar between

October 2nd to 4th, 1993. **In** *Biol. Control Ins. Pests* (Ed. S.C. Goel) U.P. Zoological Society, Muzaffarnagar. pp. 87-90.

Ballal, C.R. and Singh, S.P. 2001. Comparative studies on pheromone trap catches of *Helicoverpa armigera* (Hubner) in sunflower and pigeonpea ecosystems. pp. 177-179, In *Biocontrol Based Pest Management for Quality Crop Protection in the Current Millennium*, (Eds. Darshan Singh, M.S. Mahal, A.S. Sohi, V.K. Dilawari, K.S. Brar and S.P. Singh), 224 pp.

Ballal, C.R. and Singh, S.P. 2003. The effectiveness of *Trichogramma chilonis* Ishii, *Trichogramma pretiosum* Riley and *Trichogramma brasiliense* (Ashmead) (Hymenoptera: Trichogrammatidae) as parasitoids of *Helicoverpa armigera* (Hubner) (Lepidoptera: Noctuidae) on sunflower (*Helianthus annuus* L.) and redgram (*Cajanus cajan* (L.) Millsp.). *Biocontrol Science and Technology*, 13: 231-240.

Ballal, C.R., Jalali, S.K., Sunil Joshi and Rao, N.S. 2001. Variations in populations of *Campoletis chlorideae* Uchida and *Helicoverpa armigera* (Hubner) from different crop ecosystems pp. 11-12, In *Biocontrol Based Pest Management for Quality Crop Protection in the Current Millennium*, (Eds. Darshan Singh, M.S. Mahal, A.S. Sohi, V.K. Dilawari, K.S. Brar and S.P. Singh), 224 pp.

Ballal, C.R. Srinivasan, R. and Jalali S.K. 2009. Evaluation of an endosulfan tolerant strain of *Trichogramma chilonis* on cotton. *Biocontrol* (online version DOI 10.1007/s10526-009-9222-0).

Canard, M.,Y. Semeria and T.R. New (Eds.) 1984. Biology of Chrysopidae, Dr W. Junk Publishers, The Hague, Netherlands. 294 pp.

Chandramohan, N. 1994. Seasonal Incidence of Diamondback moth, *Plutella xylostella* L. and its Parasitoids in Nilgiris. *Journal of Biological Control*, **8**: 77-80.

Dodia, J.F., Yadav, D.N. and Patel, R.C. 1986. Management of fruit sucking moth, *Othreis fullonica* L. (Lepidoptera : Noctuidae) in citrus orchards at Anand (Gujarat) *GAU Research Journal*, **11**: 72-73.

Grafton-Cardwell, E.E. and Hoy, M.A. 1985. Intraspecific variability in response to pesticides in the common green lacewing, *Chrysoperla carnea* (Stephens) (Neuroptera: Chrysopidae). *Hilgardia*, 53: 6: 1-31.

Hoy, M. A. 1994. Insect molecular genetics: An introduction to principles and applications. insect molecular genetics, Academic Press, San Diego, California, USA, 546pp.

Jalali, S.K. and Singh, S.P. 1990. A new record of *Ooencyrtus papilionis* (Hymenoptera: Encyrtidae) on the eggs of *Papilio demoleus* (Linn.) from India. *Journal of Biological Control*. **4**: 59-60.

Jalali S.K., Singh S.P., Venkatesan T., Murthy K.S. and Lalitha Y. 2006. Development of endosulfan tolerant strain of an egg parasitoid *Trichogramma chilonis* Ishii (Hymenoptera: Trichogrammatidae). *Indian J Exp Biol* 44: 584-590.

Jayarathnam, K. 1977. Studies on the population dynamics of the diamond-back moth *Plutella xylostella* (Linnaeus) (Lepidoptera : Yponomeutidae) and crop loss due to the pest on cabbage. *Ph.D. thesis, UAS, Bangalore,* 215 pp.

Joshi, S. and Viraktamath, C.A. 2004. The sugarcane woolly aphid, *Ceratovacuna lanigera* Zehntner (Hemiptera: Aphididae): its biology, pest status and control. *Current Science* 87: 307-316.

Krishnamoorthy, A. and Singh, S.P. 1988. Observational studies on the occurrence of parasitoids of *Papilio* spp in citrus. *India Journal of Plant Protection,* 16: 79-81.

La Salle, J. 1993. Parasitic hymenopters, biological control and bio-diversity. In: La Salle J. and Gauld I.D. (eds) Hymenoptera and Biodiversity, CAB International, Wallingford, pp. 197-215.

La Salle J. 1999. Insect Bio-diversity in Agro-ecosystems: Function, value and optimization. Pp 155-182 In Agro-biodiversity: Characterisation, utilization and management. (Eds. Wood, D. and Lenne, J.M., CABI Publishing 490 pp.

Lockwood, J. A. and De Brey, L.D. 1990. A solution for the sudden and unexplained extinction of the Rocky Mountain Grasshopper (Orthoptera: Acridiidae), *Environmental Entomology,* 19: 1194-1205.

Loxdale, H.D. 1994. Isozyme and protein profiles of insects of agricultural and horticultural importance. Pp. 337–375 in Hawks worth, D.L. (*Ed.*) *Identification and characterization of pest organisms.* Wallingford, CAB International.

Nagakatti, S. and Jayanth, K.P. 1982. Population dynamics of major pests of cabbage and their natural enemies in Bangalore district (India). In *Proceedings of the International Conference of Plant Protection of Tropics,* Malaysian Plant Protection Society, Kula Lumpur, Malasyia. pp. 325-347.

Nagarkatti, S., Nagaraja, H. 1977. Biosystematics of *Trichogramma* and *Trichogrammatoidea. Annual Review of Entomology,* 22: 177-192.

Pawar, A.D. 1986. Role of natural enemies in the integrated pest management of rice. *Plant Protection Bulletin, India,* 38: 15-26.

Platner, G.R., Velten, R.K., Planoutene, M. and Pinto, J.D. 1999. Slide-mounting techniques for *Trichogramma* (Trichogrammatidae) and other minute parasitic Hymenoptera. *Entomological News,* 110: 56-64.

PDBC 2009. Annual Report of the AICRP on Biological Control of Crop Pests and Weeds – 2008-09. Project Directorate of Biological Control, Bangalore, 279 pp.

Rabindra, R.J., Mohanraj, P., Poorani, J., Jalali, S.K., Joshi, S. and Ramani, S. 2002. *Ceratovacuna lanigera* Zehntner (Homoptera: Aphididae), a serious pest of sugarcane in Maharashtra and attempts at its management by biological means. *Journal of Biological Control,* **16,** 171-172.

Reid, W.V. and Miller, K.R. 1989. *Keeping Options Alive: The Scientific Basis for conserving Biodiversity*, World Resources Institute, Washington.

Romeis, J., Shanower, T.G. and Zebitz, C.P.W. 1998. Physical and chemical plant characters inhibiting the searching behaviour of *Trichogramma chilonis*. *Entomologia Experimentalis et Applicata*, 87: 275-284.

Romeis, J., Shanower, T.G. and Zebitz, C.P.W. 1999a. *Trichogramma* egg parasitism of Helicoverpa armigera on pigeonpea and sorghum in southern India. *Entomologia Experimentalis et Applicata*, **90**: 69-81.

Romeis, J., Shanower, T.G. and Zebitz, C.P.W. 1999b. Why *Trichogramma* (Hymenoptera: Trichogrammatidae) egg parasitoids of *Helicoverpa armigera* (Lepidoptera: Noctuidae) fail on chickpea. *Bulletin of Entomological Research,* **89**: 89-95.

Singh, S.P. 2002. Progress and potential of Biological Control in India. *Thirteenth Dr. C.P. Alexander Memorial Lecture 2001*, Dept of Zoology, University of Delhi, Delhi, 32 pp.

Solbrig, O.T. 1991. *From Genes to Ecosystems: A Research Agenda for Biodiversity*. International Union of Biological Sciences, Paris.

Stouthamer, R. 1997. *Wolbachia* induced parthenogenesis, 102-124. In S.C. O'Neil, J. Werren, A.A. Hoffman (eds.), Influencial passengers, inherited microorganisms and arthropod reproduction. Oxford University Press, 226p.

Stouthamer, R. & J.H. Werren. 1993. Microbes associated with parthenogenesis in wasps of the genus *Trichogramma*. J. Invertebr. Pathol. 61: 6-9.

Subba Rao, B. R., Sangwar, H. S. Abbasi, O. A., Singh, Y. and Ksheer Sagar, A. M. 1965. New records of hymenopterous parasites of *Nipaecoccus vastator* (Maskell) (Homoptera : Coccidae), a serious pest of *Citrus* spp. from India. *Indian Journal of Entomology*, **27**: 109-110.

Tripathi, G. M. 1995. Record of parasitoid and predator complex of sugarcane woolly aphid, *Ceratovacuna lanigera* Zehntner in Nagaland. *Indian Sugar*, **44**: 883-885.

Turnock, W.J., Taylor, K.L., Schroder, D. and Dahlsten, D.L. 1976. Biological control of pests of coniferous forests. In: Huffaker, C.B. and Messenger, P.S. (Eds.) Theory and Practice of Biological Control. Academic Press, New York pp. 289-311.

Unruh, T.R. and Messing, R.H. 1993. Intra-specific bio-diversity in hymenoptera: implications for conservation and biological control. pp. 27-52. In: La Salle, J. and Gauld, I.D. (Eds.) Hymenoptera and Bio-diversity CAB International, Wallingford.

Varshney, R.K. 1998. Faunal diversity in India: Insecta, pp. 145-157. **In**: *Faunal diversity in India: a commemorative volume in the 50th year of India's independence* (Eds. Alfred, J.R.B.; Das, A.K.; Sanyal, A.K.). Zoological Survey of India, Calcutta. viii + 497 p.

Vavre F; Fleury F; Varaldi J; Fouillet P. and Boulétreau 2000. Evidence for female mortality in Wolbachia mediated cytoplasmic incompatibility in haplo-diploid insects: epidemiologic and evolutionary consequences. *Evolution; International Journal of Organic Evolution* 54(1):191-200.

Venkatesan, T., Singh, S.P. and Jalali, S.K. 2000. Rearing of *Chrysoperla carnea* (Neuroptera:Chrysopidae) on semi-synthetic diet and its predatory efficiency against cotton pests. *Entomon* 25 (2): 81-89.

Venkatesan, T. Poorani, J., Jalali, S.K., Srinivasamurthy, K., Ashok kumar, G., Lalitha, Y. and Rajeshwari, R. 2008. Confirmation of the occurrence of *Chrysoperla zastrowi arabica* Henry *et al.,* (Neuroptera: Chrysopidae) in India. *Journal of Biological Control,* 22 (1): 143-147.

Venkatesan, T., Jalali, S.K., Srinivasamurthy, K., Rabindra, R.J. and Lalitha, Y. 2009. Occurrence of insecticide resistance in field populations of *Chrysoperla zastrowi arabica* (Neuroptera: Chrysopidae) in India. *Indian Journal of Agricultural Sciences,* 79 (11) (in press).

Yadav, D.N., Patel, R.C. and Manjunath, T.M. 1975. Seasonal activity of *Apanteles plutellae* (Kurdj.), a larval parasite of *Plutella xylostella* (L.) at Anand (Gujarat, *In* Occurrence of insecticide resistance in field populations of *Chrysoperla zastrowi arabica* (Neuroptera: Chrysopidae) in India. *Indian Journal of Plant Protection,* **3**: 111-115.

Chapter - 6

Ecological Impact and Biodiversity of Insect Pests of Teak

John Prasanth Jacob
Institute of Forest Genetics and Tree Breeding, Coimbatore-641 002

Introduction

Teak (*Tectona grandis* Linn. F.) is one of the most favoured timber species all over the world, used for a range of products and services. It is known for its strength, durability and attractive appearance. Ever increasing demand for teak timber has resulted in large scale plantations, both within and outside its range of natural distribution. Teak occurs in natural forests between 9° to 26° N latitude and 73° to 104° E longitudes which includes southern and central India, Mayanmar, Laos People's Democratic Republic and northern Thailand (White, 1991). In India there are 8.9 million hectares of teak bearing forests (Tewari, 1992) within precipitation range of 800 to 2500 mm. It grows well from sea level to an elevation of 1200m. There has been a steady expansion of teak plantation over the decades in order to meet the dwindling supply of wood from natural forests and due to an increase in demand for wood products. Teak plantations were raised to meet the industrial needs and to meet the needs of local livelihood and global environment. Such multi-functional plantations should minimize insect pest and disease risk and should provide better environment for integrated pest and disease management. Among other plantation tree species teak occupies a prominent place and there is and emerging concern about the ecological impact and biodiversity of insect pests of teak.

Insect Pest Diversity of Teak

About 187 species of insects are reported to be associated with living Teak. Many of these insects are minor or occasional pests and about 6 of them are found to cause considerable damage.

They include the white grubs which attack the seedlings in the nurseries; the defoliator *Hyblaea puera* (Lepidoptera : Hyblaeidae), and skeletoniser *Eutectona machaeralis* (Lepidoptera : Pyralidae) which defoliate the plant during all stages; *Sahyadrassus malabaricus* (Lepidoptera : Hepalidae) and *Indarbela quadrinotata* (Lepidoptera : Metarbeliadae), boring into the stem, during sapling and pole stages and *Alcterogystia cadambae* (Lepidoptera: Cossidae), the 'trunk borer' endemic to certain specific localities. *Asphondyla tectonae* (Diptera : Cecidomyiidae) and another Cecidomyiid induce galls in stem and leaves respectively. Teak grass hopper *Aularches scabiosae* (Orthoptera: Acrididae) also defoliates nursery seedlings and young saplings.

Hyblaea puera

H. puera, a polyphagous pest feeds on many tree species, but teak is considered as the principal host (Plate 1A). The first and second instars feed mainly on the leaf surface. From the third instar onwards the larva cuts out a leaf flap, folds it over and fastens it with silk and feeds from within. Incidence of heavy defoliation, particularly in young plantations sometimes causes die-back of the leading shoots, resulting in epicromic branching and further loss of quality. Though the pest is found throughout the year in the forest areas, its population reaches peak during pre-monsoon showers. The moths when reach a critical density in a particular location migrate to facilitate exploitation of new food sources (Nair *et al.,* 1985).

Eutectina machaeralis

E. machaeralis is known as the leaf skeletoniser to teak. Feeding superficially to begin with, the larvae later consume the leaf tissues between the networks of veins, thereby skeletonising the leaves (Plate 1B). The larvae generally feed on mature, tough leaves, unlike *H. puera,* but when given the choice, they will select younger and softer leaves on which the rate of growth is faster (Beeson, 1941). The defoliation caused by the skeletoniser is not as serious as that of *H. puera.* The epidemic of the skeletoniser is not a regular annual phenomenon, unlike that of *H. puera.* Generally, the pest infestation occurs from Octorber to May. Severe attack of the skeletoniser is reported to cause die-back of leading shoots in young plantations which result in undesirable forking of saplings. Epidemic out breaks are common in central India.

White grubs

Teak seedlings are attacked in the nurseries by the larvae of some scarabaeid beetles known as white grubs, especially belonging to the genus *Holotrichia*. Grubs are fleshy, cream coloured and C-shaped and feed on the roots of seedlings. The infested seedlings show symptom of wilting and eventually they succumb to attack. Mortality up to 80% is reported to occur due to the infestation of white grubs in nurseries.

Termites

Teak is not normally attacked by termites, but the plantations raised in the semi-arid areas of Tamil Nadu are found severely infested by them. The termites eat away both the inner as well as outer bark of saplings in patches, up to about $3/4^{th}$ their height, exposing the wood. The attack may leads to invasion of pathogens into the stem, desiccation and resultant stress.

Aphids

Nurseries and young plantations raised near agricultural fields suffer the attack of the Cotton Aphid, *Aphis gossypii*. Both the adults and nymphs suck sap from the lower side of leaves and other tender parts. The infested leaves show symptoms of curling. They shrivel, dry and fall off prematurely. The large amounts of honey dew secreted by the pest promote the growth of sooty moulds.

Sahyadrassus malabaricus

S. malabaricus is a polophagous pest which attacks the saplings. Its infestation is confined to penisular India, in states like Kerala, Karnataka, Maharashtra and Tamil Nadu. The young larvae appear to develop on ground vegetation, before they migrate to teak saplings. On entering into the stem of sapling, the larva excavates a long cylindrical tunnel along the pith, sometimes extending to the tap root. The top end of the tunnel is curved and opens to outside. The mouth of the tunnel is covered by a mat work consisting of coarse saw dust like particles of wood and bark, spun together with silk secreted by the larva. The large mat converging made by larva is an indication of the pest attack. The larva feeds on the callus tissue which develops around the tunnel moth due to constant feeding. The larva feeds during night under the mat cover. The sapling sometimes may break at this attacked weak portion.

Indarbela quadrinotata

The bark feeding caterpillar, *I. quadrinotata* is also a polyphagous pest (Plate 1C). The larva enters into the stem by making a short tunnel, usually at the junction of a dead branch or snag and the bole. The tunnel is used by the larva as a shelter and it comes out during night to eat on the outer surface of the bark, excavating broad irregular patches and paths, which are of varying lengths. These areas are roofed with silk and fragments of bark and the faecal pellets. Severe attack of the bark feeding borer may lead to the invasion of pathogenic microorganisms into the saplings and they become easily prone to diseases.

Alcterogystia cadambae (Cossus cadambae)

A. cadambae is known as the "trunk borer" relatively attacking the trees at the age of 15-20 years. The insect has assumed major pest status in certain localities of Kerala, Tamil Nadu and Karnataka. The young larvae hatching out from the eggs establish in the axis of side shoots in crevices or injuries on the bark or in sites of earlier infestation. The larvae as they grown become dark red colour. The larval feeding results in the girdling of the side shoots which leads to die-back and is one of the initial symptoms of attack. The larva feeds on the bark or callus tissues, including the outer bark. It makes a tunnel of about 6-7 cm, in the sapwood. During rains, water enters into the stem through the tunnels made by the larvae, which provides a conducive condition for the growth of certain species of saprophytic fungi (Mathew and Rugmini 1996). Due to the action of such fungi, the heart wood gets rotten and ultimately the wood becomes useless.

Tetranychus urticae (Tetranychidae)

This is an occasional pest occurring in nursery seedlings (Plate-1D). The caramine red coloured mites occur underneath the leaves of teak saplings feeding on the plant sap. This results in chlorotic spots which coalesce into pale patches. There will be extensive webbing underneath the leaveas. Leaves start drying from the edges and slowly wither away. The most notable pest of teak in natural forests, plantation and in forest nursery is the defoliator *H. puera* (Jacob, 2008).

Plate 1 Insects associated with Teak A – Larva of *Hyblaea puera*, B – Larva of *Eutectona machaeralis*, C- Larva of *Indarbela quadrinotata*, D- Adult mite *Tetranychus urticae*, E – *Myllocerus dorsatus*, F- *Dereodus* sp. G - Hairy Caterpillar, H - Leaf gall by Cecidomyiid.

Plate 2. Insects associated with Teak A – Adult grass hopper *Aularches scabiosae,* B – Larva of *Pagida salvalis,* C – Larva of Buzura sp.

Asphondyla tectonae

The twig gall midge, *Asphondyla tectonae,* noticed in the natural forest in kerala and Karnataka in southern India, has also been found in poor class teak forests in central India (Beeson, 1941). It attacks the new shoots of teak and causes formation of globular, multilocular galls that coalesce, harden and surround the stem of twigs.

Aularches scabiosae

Commonly known as the spotted locust is an arboricole form found distributed in the teak forests. Nymphs are typically gregarious but not the adults. Nymphs and adults defoliate teak trees (Plate 2 A). A close synchronization of the life cycle of *A.*

scabiosae with the phenology of *T. grandis* is peculiar in which the adults lay egg in the soil as soon as the teak tree start leaf shedding during winter months. The young one will emerge only when the teak put forth new flush during monsoon showers during the following year (Jacob, 1991).

Flower and fruit feeders

The larvae of the pyralid *Pagyda salvalis* (Dabral and Amin 1975) and the Geometrid *Buzura* sp. (Jacob unpublished) cause severe damage to flower buds and young fruits in the inflorescence. Incidence of *Buzura* sp. was more in Teak seed orchards. *P. salvalis* bores in to the flower buds and fruits. Late stages of *Buzura* sp. feed the whole buds and fruits (Plate 2 B & C). Up to 28% of bud loss during flowering by *P. salvalis* and 37% of fruit loss by *Buzura* sp. was observed in teak inflorescence. A small population of 4-5 individuals of a tiny cerambycid beetle feed on the dehisced pollen on the stamens.

Other common pests

***Myllocerus* sp. (Coleoptera : Curculionidae)**

The weevil is a persistent pest in nurseries and plantations. Adults usually hide below the leaf and feed on the leaf margins. During severe defoliation all the leaves have notched margins resulting in wilting of plants in patches (Plate 1 E & F).

Hairy Caterpillar

An occasional pest with epidemic out break in plantation. The hairy caterpillar feeds on leaves indiscriminately (Plate 1G).

Cecidomyiid

A species of cecidomyiid midge produces puff like galls on the under surface of the leaves. Occasionally the population builds up and the leaves dry up. There is little information on the biology of this species (Plate 1 H).

***Hyblaea puera* – Teak interaction**

Teak is one of the most extensively studied timber species for its insect pest problems. The most notable pest of teak is the teak defoliator in natural forests, plantations and in forest nursery. Therefore, lot of research work has been carried out on the defoliation related loss of wood increment, population movement, and various pest management measures. Outbreaks of this major pest results in complete defoliation of teak stands, resulting in reduced growth, poor form and occasionally tree mortality. Teak defoliator outbreaks are regular annual phenomena in Kerala. Outbreaks begin in comparatively small epicenters in large teak plantations, and are associated with pre-monsoon showers usually in April. In general two to three major outbreaks occur during the infestation period. Early detection of the epicenter build up or of a full outbreak is very difficult because the epicenter locations are not constant over the years. Similarly the outbreak locations are also not constant as the moths move away from the emerging sites as part of their short-range migratory behaviour. Another major problem in population detection is that the damage on the tree is visible from ground only when the second instar larvae cut the leaf and make folds.

Studies carried out at the Kerala Forest Research Institute (KFRI) have indicated losses averaging 44 percent of potential growth increment in four-to eight-years-old trees due to the teak defoliator (Nair *et al.*, 1985; Nair *et al.*, 1996). Estimation of growth losses and economic impact due to the teak defoliator, *H. puera* has been made in the past (Mackenzie, 1921 – loss of 6.6% of potential growth; Beeson, 1941 – loss of 13% of Mean Annual Increment). Nair *et al.* (1985) based on a statistically designed study showed that in a young teak plantation, average of 44 per cent growth loss (measured as volume increment) occurred for the five years of experiment. In other words, trees protected against *H. puera* put forth an annual increment of 6.7 m^3 ha^{-1} compared to the mean annual increment of 3.7m^3 ha^{-1} of unprotected trees, which amounts to a gain of 3m^3 ha^{-1} annually. Theoretically the protected trees can be expected to yield the same volume of wood in 26 years as unprotected trees would yield in 60 years. *H. puera* can sometimes cause more than one total defoliation per year, thus the intensity of growth loss will be more. In younger plantation, dieback of the leading shoot results in forking and poor growth form.

In order to demonstrate the need for control against *H. puera,* permanent demonstration plot has been maintained at Nilambur, Kerala. Plots were left to natural defoliation and in the other plots appropriate measures were adopted to control the teak defoliator as and when required. Growth data for five years from these plots showed that in the protected plot there was 45 per cent increase in mean height and 19% increase in girth at breast height over the control plot (Varma *et al.*, 1998).

Management Measures

Control measures against the teak defoliator has not become part of the management practice, at least in the government owned plantations. Achievement of practical control depends on our ability to predict or detect the outbreaks promptly through a surveillance system and initiate action with the speed and efficiency similar to that of a fire-fighting organization. Pesticide application is common in forest nurseries. Environmental considerations and the cost involved limits pesticidal application in plantations.

The Diversity of Parasitoids

The total number of parasitoids recorded from *H. puera* from the Indian region stands at 40, of which 24 have been recorded from Nilambur, Kerala at one time or another -10 from larvae and 14 from pupae. Regional differences within India in the distribution of *H. puera* parasitoids have been noted earlier (Beeson, 1934). It is obvious that parasitoid populations are unable to respond numerically to host density increase because of the highly mobile nature of *H. puera* populations. When a new generation of adult parasitoids is built-up in a locality, the host moves away to another area. Similarly when a high-density population of the susceptible stage of host larvae is built up from an incoming population of moths, there is rarely a resident population of parasitoids. This results in spatial separation of the host and its parasitoids, host migration helping it to evade the parasitoids (Nair, 1987). This explains the lack of numerical response of the parasitoids to host density. Under these circumstances, natural enemies locally conserved by the silvicultural methods recommended earlier cannot prevent outbreaks caused by an incoming wave of moth population. The circumstances leading to the first outbreak during the early part of the teak growth season which then spreads out like a chain reaction remains largely unknown, but evidences suggest the role of characteristic moth aggregations, and wind-aided concentration and displacement of airborne moths. It is therefore unlikely that locally-conserved natural enemies can prevent the development of outbreak epicenters.

Aggregation of flocks of crows often found in teak plantations during the later stages of defoliator outbreaks. Such behavioural mechanisms might be operating in some insect parasitoids as well. By aggregating in certain patches, particularly of high host density, the parasitoids are leaving other patches as partial refuges from parasitism (Waage and Hassel, 1982), thereby ensuring the survival of both the host and the parasitoid. The same situation arises due to the migration of *H. puera* moths (Nair 1987). The progeny of the incoming moths have little pressure from endemic natural enemies, and the successfully emerging parasitoid adults cannot find sufficient hosts to oviposit. Inoculative release of exotic parasitoids is not applicable against *H. puera* as the pest is native. The most promising species for the applied biological control of *H. puera* through inundative release of mass reared parasiotoids are the bethlid *Goniozus montanus* and the eulophids *Sympiesis* sp. and *Elasmus hyblaeae.* The usefulness of the braconid *Apanteles machaeralis* and the ichneumonid *Eriborus gardneri* needs further study.

The sudden occurrence of outbreaks, the short duration of the susceptible early larval instars of the host and synchronous development of the host during outbreaks (i.e., uniform age distribution of the population) makes the proper timing of parasitoid release a crucial factor for success.

The use of parasitoids as a vector or dispersal agent for biocontrol agents like *Bacillus thuringiensis* and the specific *H. puera* NPV is also being explored.

Microbial Control

The *H. puera* NPV (HpNPV) is a DNA virus belonging to the family Baculoviridae. Preliminary studies indicated that HpNPV could be propagated on *H. puera* larvae in the laboratory and that a high volume application of a crude, unformulated preparation applied against the pest in the field could reduce the leaf damage up to 76% (Nair *et al.*, 1998). Compared to many other baculovirues, the HpNPV is quick acting and causes significant mortality of the larvae in about three days. However, studies also indicated that when applied in the field, the virus had only low persistence.

Microbial agents have proved to be a viable alternative to chemical insecticides, especially in the management of forest insect pests in different parts of the world. In India, little attempt has been made to explore the feasibility of using microbial agents in forest pest management. Intensive research on teak defoliator management was started in the early 1990s, following the detection of a baculovirus from Kerala. Preliminary studies demonstrated the efficacy of HpNPV which lead to further advanced studies in this field. The third instar larva has been identified as the most appropriate target larval instar for virus application. Compared to the first and second instars, it consumes more foliage per unit time and, therefore, has a greater likelihood of taking in the virus dose. The height of the teak tree is always a limiting factor in achieving successful control of the pest. Further trials using other ultra low volumes sprayers that can carry spray droplets to greater height up the canopy need to be undertaken in order to extend the control operations to taller trees.

Host Plant Resistance for Teak Defoliator

Selection of pest resistance tree from wild populations is another choice for dealing with pest problems. It is an ecological approach to make the environment unfavourable to pests and less harmful to the natural enemies is needed. There are innumerable possibilities for an unfavorable condition to exist in a plant that would reduce the possibility of its successful utilization by an insect. This unfavourable condition that could be of chemical or physical basis adversely affects the insect's feeding, growth and reproduction leading to insect pest resistance in plants. Limited works have been carried out to identify and evaluate resistant phenotypes in teak. Identification of a resistant crop / tree variety, which forms the basic foundation for an integrated pest management programme, holds promise for effective suppression of defoliators in teak. Meanwhile, clonal forestry is now well recognized as a method for mass production of desired trees to get maximum economic benefits since it enables the capture of genetic gain generated by skillful selection. During the last two decades clonal planting of species like teak, poplar, eucalyptus and casuarinas is being extensively carried out commercially under agroforestry system as well as in many other afforestation programmes. Clones derived from different provenances or land races of teak showed variation in levels of defoliation by *H. puera* (Jacob *et al.,* 2002) and *E machaeralis* (Roychoudhury *et al.,* 1997). While comparing the incidence and extend of damage by teak defoliator on various teak clones, a significant difference was observed in the extent of feeding damage on certain clones of teak in the field.

Distinct morphological, physiological and biochemical differences are found in association with different levels of susceptibility, tolerance and resistance. These variations have been documented and methods to breed for resistance are programmed. Comparison of feeding area by *H. puera* on teak clone leaves under no choice and choice tests showed significant variation in the extent of feeding by *H. puera* among the clones (Jacob and Balu, 2007). Observations on the feeding behaviour of *H. puera* larvae showed that feeding was mostly attempted at the margins and patches of veins were left in the lamina with only the tissues scrapped and fed in resistant clones while the whole leaf is consumed in susceptible clone (Plate 3).

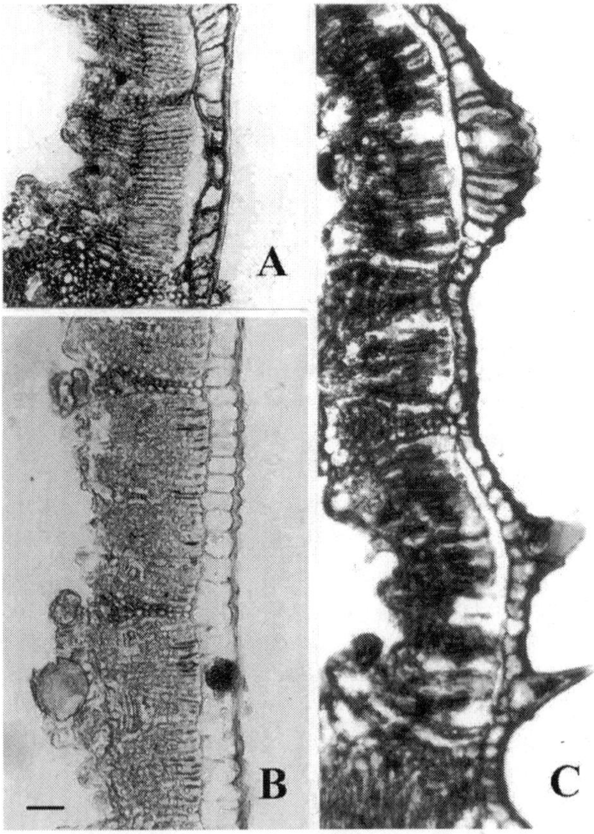

Plate 3. Cross section of Teak clone leaves — **A.** Susceptible clone leaf; **B.** Moderately susceptible clone leaf; **C.** Resistant clone leaf.
(Bar represents 40μ)

Histological studies of leaves showed marked variation in leaf surface cuticular layer among clones. Most susceptible clones have a smooth or even cuticular layer whereas in resistant clones it is rough. The adaxial surface of resistant clones is rough with numerous convex projections topped with a single, pointed, hard trichome. In addition to this numerous long, sharp and short, blunt spines are also distributed on the adaxial surface. The abaxial surface of unattacked clones also has numerous long and short soft hairs with a density of 5460-5500 sq. cm. In the case of most susceptible clone, the upper cuticular surface is smooth without any spines. Very few diffused type of cuticular thickenings of only 100 microns in diameter were observed. Lower surface of these clones had isolated, multi-cellular long hairs.

Table 1. Phenols, flavanoids and anthocyanidins in Teak clones

CLONES	PHENOLS*	FLAVANOIDS*	ANTHOCYANIDINS*
Susceptible clone	p-Hydroxybenzoic acid Vanillic acid Resorcinol	Naringenin (Flavanone) Apigenin (Flavone)	Pellargonidin Delphinidin
Moderately susceptible clone	Vanillic acid Orcinol Salicylic acid Resorcinol	Phloridzin (Dihydrochalcone) Naringenin (Flavanone) Naringin, Hesperidin (Favanone)	Anthocyanin groups were expressed in the chromatogram but not specified by any anthocyanidins.
Resistant clones	Orcinol Protocatecheuic acid Pyrogallol Salicylic acid 2-Methyl resorcinol	Aureusidin 6-glucoside Aureusidin 4-glucoside	Anthocyanin groups were expressed in the chromatogram but not specified by any anthocyanidins
	Oricinol Pyrogallol	Hesperidin	Anthocyanin groups were expressed in the chromatogram but not specified by any anthocyanidins
	Oricinol Pyrogallol	Aureusidin 4 glucoside Naringin	Anthocyanin groups were expressed in the chromatogram but not specified by any anthocyanidins

* Fractions with best separations were eluted

The total protein content of resistant clones was comparatively lesser than the clones of most and moderately susceptible category. Further, the total carbohydrate and lipid contents in the leaves of unattacked or resistant group were higher than the most and moderately susceptible clones. Phenols such as Orcinol, and Pyrogallol were found common in the clones of unattacked category as compared to the susceptible clones, where these compounds were either absent or in trace amounts (Table 1). Protocatechuic acid and 2-Methyl resorcinol were also found in resistant clones. Hesperidin, naringenin and naringin formed the common flavanoids in all the tested clones.

Pollinators of teak inflorescence

When there is a demand to improve genetic resources of timber species like teak, low fruit production by teak impedes planting programme with genetically superior and improved seeds. Expected out put from clonal seed orchards are also not achieved. Though large numbers of flowers are produced by teak, only a small percentage develops in to fruits. Various biotic and abiotic factors are responsible for the low seed production in teak. Some studies have brought to light the beneficial entomofauna visiting teak inflorescence (Egenti, 1974; Mathew *et al.*, 1987; Sasidahran and Nagarajan, 2005; Jacob and Nicodemus 2007 unpublished) (Table 2, Fig. 1) besides birds (Nicodemus and Jacob, 2004; Nicodemus *et al.*, 2005).

Table 2. Insect visitors to teak inflorescence and the pollen load.

No	Species	No. of species	Pollen count (X1000 grains ml^{-1})
HONEY BEES			
1	*Halictus* sp. (Family:Halictinae)	3	96
2	*Hylaeus* sp. (Family:Collectidae)	1	3.33
3	*Trigona iridipennis* (Family:Apidae)	1	2
4	*Braunsapis* sp. (Family:Anthophoridae)	2	1.66
WASPS			
1	*Scolia* sp. (Family: Scoliidae)	2	0.4
BUTTERFLIES			
1	*Hypolimnas boline*	1	0
2	*Tros hector*	1	0
3	*Atella phalanthor*	1	0
4	*Danais limniace*	1	0
5	*Euploea coreta*	1	0
6	*Paplio budha*	1	0
BEETLES			
1	Cetoniinae (Family : Scarabaeidae)	1	0.75
FLIES			
1	*Stratiomyiidae* fly	3	0.5

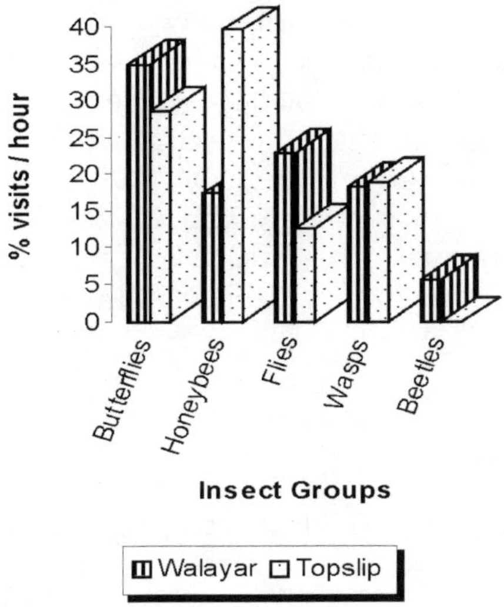

Fig. 1. Visitation by different insect groups on Teak inflorescence

The diversity of entomofaunal visitors of teak flowers does not differ in plantations and natural populations (Hedegart, 1976). Total number of species reported so far ranged from 17 to 38 and those carrying pollen on their bodies was 5 to 15. Hymenopterans (bees and wasps) were the dominant group among the visitors constituting 50% of all visitors (Tangmitcharoen and Owens, 1997). Seed orchard environments are similar to other teak populations in terms of resource availability for potential pollinators and lack of which cannot be a reason for low fruit and/or seed set in orchards. Three species of *Halictus* were found actively feeding nectar and collecting pollen. Solitary bees viz., *Prosopis pratensis*, *Halictus* sp. and *Allodope marginata* were found to be abundant in numbers and frequent in visits (Mathew *et al.,* 1987; Jacob and Nicodemus 2007-unpublished). Visitation of the insects was observed to be mostly within a tree and they usually moved to another inflorescence in the same rather than to another tree. Honeybees dominate in their visitation rates also. Added to this is the large quantity of teak pollen carried by individuals, which contribute to progeny quality through out crossing in an environment like seed orchards (Quesada *et al.,* 1993, Keams *et al.,* 1998). More trees in flower at a given time provide habitat connectivity which will maintain abundant and diverse pollinators and also plant-pollinator interactions in teak clonal seed orchards (Dewenter and Tscharntke, 1999). Even though *T. iridipennis*, *Hylaeus* sp.

Braunsapis sp. carried considerably less amount of pollen, their visiting frequency was more than that of wasps and beetles. Though butterflies also make considerable number of visits, absence of pollen on their body parts make them an inefficient pollinator of teak flowers (Fig. 2). Multiple visits and visiting more flowers per visit and presence of teak pollen alone on their bodies is likely due to the presence of teak flowers as the only source of food in the vicinity (Willis and Kevan, 1995).

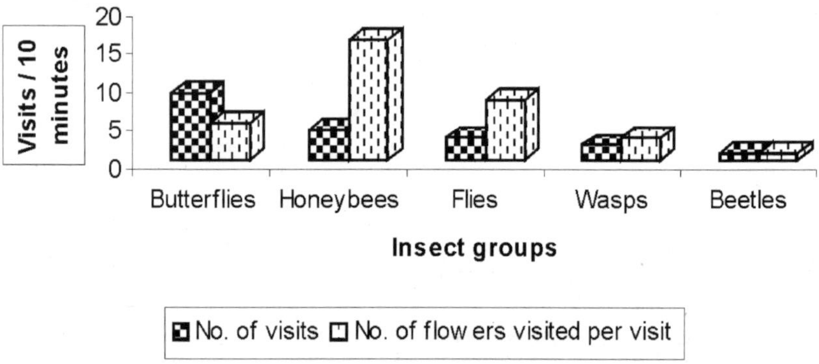

Fig 2. Average visits made by different insect groups to Teak inflorescence

Conclusion

Plantation expansion is a factor favouring pest outbreaks. Outbreak development of *H. puera* is intriguing and its dynamics are not fully understood. There is dearth of data on the outbreak dynamics of other insect key pest species like the skeletoniser *E. machaeralis,* the flower and fruit feeder *P. salvalis* particularly in central India and the borer *A. cadambae* in Southern peninsula. All these species cause damage in natural forests and plantations. However outbreaks are conspicuous and appear to be more frequent in plantation. There are ecological and environmental limitations in employing management methods involving pesticide application and biological control through natural enemies. Raising of few clones of teak should be avoided in plantation programmes. Pest resistant or tolerant as well as productive germplasm/clone for should be identified in teak for future plantation programme. In future, care may be taken to incorporate appropriate genetic diversity with periodic introductions and phasing out of different clones or provenances in plantation programs to minimize economic damage due to pest attack.

References

Beeson, C.F.C. 1941. The ecology and control of the forest insects of India and the neighbouring countries. 1993 reprint. Bishen Singh Mahendra Pal Singh, Dehra Dun, India. 1007p.

Dabral, S.L. and Amin, P.W. 1975. Poor fruit formation in Teak in chanda forests of Maharashtra. *Indian Forester*. 616-619.

Dewenter, I.S. and Tscharntke, T. 1999 Effects of habitat isolation on pollinator communities and seed set. *Oecologia* 121: 432–440.

Egenti, L.C. 1974. Preliminary studies on pollinators of Teak (*Tectona grandis* L.f.) Research Paper No. 29. Research Paper Forest Services, Federal Department of Forest Department of Forest Research, Nigeria. 240. Tangmitcharoen and Owens D *Biology, Pollination, Receptivity and pollen Tube Growth of Teak*.

Hedegart, T. 1976. Breeding system, variation and genetic improvement of teak (*Tectona grandis* Linn. f). In: Burley, J. and Styles, B.T. (Eds), *Tropical Trees*, Linnean Society Symposium Serires No. 2 Academic Press, London. pp. 109-123.

Jacob P.J. 1991. Insect Host Plant Interactions with special reference to the nutritional and reproductive aspects of *Aularches scabiosae* F (Insecta: Orthoptera: Acrididae) on *Tectona grandis* Linn F (Verbenaceae). Ph. D Thesis, University of Madras, Chenani.

Jacob, J.P. 2008: *Foresters Manual on Nursery Pests and Their Management*. (ICFRE Publication No. 126 ICFRE Bk 73/IFGTB Bk-1 2008).

Jacob J.P., A. Balu, S. Murugesan, B. Deeparaj and G. Srinivasan 2002. Variation in defoliator incidence on teak clones. *Proceedings of the National Symposium- Vistas of Entomological Research for the new Millennium* (Eds) K.P.p. 164-172.

Jacob J.P. and A. Balu 2007. Biochemical and cuticular variation in teak clone leaves and resistance to teak defoliator. *Indian Forester* 133: 197-205.

Mackenzie J.M.D. 1921 Some notes on forest insect pests in Burma. *Indian Forester* 47: 309-317.

Mathew, G. and Rugmini, P. 1996. Impact of the borer *Alcterogystia cadambae* (Moore) (Lepidoptera: Cossidae) in forest plantations of teak in Kerala, India. *In*: Nair, K.S.S., Sharma, J.K. and Varma, R.V. (eds) *Impact of diseases and insect pests in tropical Forests*, 304-310. Proceedings of the IUFRO Symposium, Peechi, India, 23-26. November 1993. Kerala Forest Research Institute, Peechi, India and FAO/FORSPA, Bangkok.

Mathew, G., Koshy, M.P. and Mohandas, K. 1987. Preliminary studies on insect visitors to teak (*Tectona grandis* Linn. F) inflorescence in Kerala, India. *Indian Forester* : 61- 63.

Nair, K.S.S., Sudheendrakumr, V.V., Varma, R.V. and Chacko, K.C. 1985. Studies on the seasonal incidence of defoliators and the effect of defoliation on volume increment of teak. *KFRI Research Report No.* 30, Kerala Forest Research Institute, Peechi, Kerala. 78p.

Nair, K.S.S., Sudheendrakumar, V.V., Varma, R.V., Chacko, K.C. and Jayaraman, K. 1996. Effect of defoliation by *Hyblaea puera* and *Eutectona machaeralis* (Lepidoptera) on volume of increment of teak. *In*: Nair, K.S.S., Sharma, J.K. and Varma, R.V. (eds) *Impact of diseases and insect pests in tropical forests*, 257-273. Proceedings of the IFURO Symposium, Peechi, India, 23-26 November 1993. Kerala Forest Research Institute, Peechi, India and FAO/FORSPA, Bangkok.

Nair, K.S.S. 1987. Life history, ecology and pest status of the sapling borer, *Sahyadrassus malabaricus* (Lepidoptera, Hepialidae). *Entomon* 12: 167: 173.

Nicodemus A. and J. Prasanth Jacob 2004. Bird pollinators of Teak *News Letter for Bird Watchers* 44: 68-69

Nicodemus, A., J. Prasanth Jacob and B. Nagarajan 2005. Pollination by nectarivorous birds in Teak clonal seed orchards. *Indian Forester*. 131: 1613-1616

Quesada, M., Winsor, J.A. and Stephenson, A.G. 1993. Effects of pollen competition on Progeny performance in a heterozygous cucurbit. *The American Naturalist* 142: 694- 706.

Roychoudhury, N, Avinash Jain, K.C. Joshi and R.B. Lal 1997. Natural resistance in teak clones to leaf skeletonizer *Eutectona machaeralis* Walker: An appraisal *Indian Forester*. 123: 1027-1035.

Sasidharan K. R. and B. Nagarajan 2005. Diversity of Hymenopteran pollinators of teak. Advances in Pollen Spore Research Vol. XXII, 59-65.

Tangmitcharoen S and J.N. Owens 1997. Floral Biology, pollination, pistal receptivity and pollen tube growth of teak (*Tectona grandis* Linn f.) *Annals of Botany*, 79. 227-241.

Tewari, D.N. 1992. *A monograph on Teak* (*Tectona grandis* L.f.). International Book Distributors, Dehra Dun, India. 479p.

White, K.J. 1991. Teak: some aspects of research and development. Publication 1991/17. FAO Regional office for Asia and the Pacific (RAPA) Bangkok.

Waage, J. K. and M.P. Hassel 1982. Parasitoids as biological control agents- a fundamental approach. *Pestology* 84: 241-268.

Willis, D.S and Kevan, P.G. 1995. Foraging dynamics of *Peponapis pruinosa* (Hymenoptera : Anthophoridae) on Pumpkin (*Cucurbita pepo*) in southern Ontario. *The Canadian Entomologist* 127 : 167-175.

Chapter - 7

Ecological Interactions and Impact of Biodiversity of Insects in Social Forests

S. Murugesan and N. Senthilkumar

Division of Bioprospecting, Institute of Forest Genetics and Tree Breeding, R.S. Puram, Coimbatore -641002, Tamil Nadu (India)

Biodiversity is often used as a measure of the health of biological systems. Tropical forest insects which dominate the number of described and estimated species indicate in general that forests harbour the highest proportion of the world's biodiversity and that the larger the forests are, the higher is the biodiversity they contain (Duelli *et al.,* 2002). Insects and plants are becoming extinct because of habitat loss, over-exploitation, pollution, over population and the threat of global climatic changes. Insects comprise the largest group of organisms and are involved in various vital 'ecosystem services' such as pollination, decomposition, herbivory and biological control as well as contributing directly to human-based economies through silk, lac and honey production (silkworm, lac insects and honey bee). Changes in the biodiversity of an ecosystem can affect ecosystem properties. If components of biodiversity are lost, the overall state and functioning of the system can be impaired (Scherber *et al.,* 2006). The role of natural hazards such as windthrow, fire, landslides, as well as that of management such as clearing, logging, and thinning have been reinterpreted from the viewpoint of disturbance ecology, biodiversity,and species conservation (Moretti *et al.,* 2004).

INSECTS IN SOCIAL FOREST STANDS

It is difficult to make an assessment of the revenue loss caused by forest insect pests in a vast and diverse country like India. One earliest estimate of annual loss due to insect damages puts down the figure as Rupees 53.70 millions (10% of the total revenue of the forests which was Rs. 537 millions in 1959-60)

(Beeson, 1941). A loss of Rs. 130/- per acre was estimated in teak *(Tectona grandis* Linn.) as a result of forking below the exploitable height of bole due to defoliators, *Hyblaea puera* Cram. and *Eutectona machaeralis* (Walker) (=*Hapalia machaeralis* Walker). Apart from the loss due to forking of the main bole, the reduction in the volume increment of timber in teak is the most serious impact of defoliation due to these pests (Nair, 2007). A careful perusal of literature clearly indicates that almost all forest tree species have been submitted to invasion of insects (Mathur and Singh, 1960). With the rapid increase in population, the pressure on fast dwindling forest resources has increased tremendously in the recent past. As a measure of conservation of biodiversity in the reserve forests, monocultures (plantations/man made forests), for subsidized supply of raw material to the industrial sector were encouraged and felling of trees and extraction of wood from reserve forests was prohibited through forest legislation (Indian Forest Act, 1927 and Forest (conservation) Act, 1980). The problem of renewing the national resources is of growing concern to many tropical countries. At the present stage of forestry development in most developing countries, forest management should generally be understood as a concept of forest resources planning in a particular area (Borota, 1991). Insect pests of various kinds play an important role in the success of man made forests. When a man made forest is created with exotic species, many pests were observed to shift from original host to introduced exotic host plants, e.g., *Coelosterna scabrator,* a root and shoot borer of *Acacia* spp. has turned out to be a serious pest on *Eucalyptus* spp. in Uttar Pradesh, Andhra Pradesh and Madhya Pradesh. *Apriona cineria* (Lamiidae: Coleoptera), a serious pest of apple trees and *Morus indica* has adapted to poplars *(Populus* spp.) as its hosts (Chatterjee and Thapa, 1964). Monocultures that are normally uniform in age and growth, support poor floristic and faunistic compositions, are highly vulnerable to serious insect attacks compared to virgin forests as they exhibit high degree of resistance to insect outbreaks. This has been observed in spruce monocultures of Central Europe, which were devastated by Spruce budworm *Choristoneura fumiferana* Clem., teak plantations of South India by defoliator complex, *Hyblaea puera* and *Hapalia machaeralis,* 'sal' plantations *(Shorea robusta)* of central India by a heartwood borer, *Hoplocerambyx spinicornis (Coleoptera: Cerambycidae), Ailanthus* plantations of South India by defoliators, *Eligma narcissus* and *Atteva fabriciella,* 'semul' plantations of (*Bombax malabaricum)* Bihar and West Bengal by a shoot borer, *Tonica niviferana,* poplar *(Populus* spp.) monocultures of Himachal Pradesh defoliated by *Pygaera cupreata* and *P. fulgurita,* (Chaterjee and Sen-Sarma, 1962) and 'Deodar' *(Cedrus deodara)* plantations of Himachal Pradesh by a geometrid defoliator, *Ectropis deodarae.* The problem of pestilent insects on forest trees is further compounded by the fact that changes have occurred over the years in forestry management practices and utilization of its products with reliance on man-made plantations. Under such altered conditions, most forest insects multiply to amazing and unmanageable levels on account of their high reproductive potential, short life cycles and continued availability of hosts. Consequently, very few practical problems concerning destructive species have been solved satisfactorily. Social forestry and many afforestation programmes in India are all part of the response to tropical deforestation and are central to much rural development. Social forestry programmes are by and large intended to be implemented in village common lands, wastelands, homesteads along the bank of ponds, canals etc. In the selection of tree species, the emphasis given on multipurpose plant species which yield fuel wood, fodder, fruits, edible and non edible oils etc. The common species planted in social forestry schemes are *Acacia, Albizia, Madhuca, Eucalyptus, Terminalia, Emblica, Gmelina, Casuarina, Moringa, Pongamia, Prosopis, Syzygium, Sesbania* and *Tamarindus* etc. All these species are grown as monoculture in order to achieve higher yield. Once monoculture of these species are introduced, naturally insect pest populations build-up rapidly on account of easy availability of quantity of host plant, less natural enemy complex and short life cycle. Therefore, it is quite natural that insect pests cause damage from nursery to young plantation and upto mature and grown up trees. Since the entire programme of social forestry and plantation forestry is meant for the relief to the rural poor, the locations selected for social forestry are site specific and ecologically diversified. In this Chapter, an account of various insects associated with social forestry, their pest status, damage caused, their natural enemies and a little bit of management practices has been dealt.

Social forestry programmes represent an effort to provide self-sufficiency and relief to the rural poor by providing them with cheap and fast growing natural resources of food, fuel, fodder, and timber. Recent agricultural practices and forest management have changed the agro- and forest ecosystems to a considerable extent. These changes paved the way for several minor pests to take the status of major ones. A total of 112 species of insects including 28 species of natural enemies were found to be associated with the nurseries and plantations of *Acacia nilotica, Tamarindus indica, Pongamia glabra, Azadirachta indica, and Albizzia lebbeck* which are grown in social forest stands. The insect-pest complexes include defoliators (48), sap-suckers (27), natural enemies (28), bark feeders (3), pod-borers (2), gall marker (1) and flower feeders (3). Among these defoliators and bark feeders were the most destructive groups. Twenty-eight species of the insects functioning as biological control agents have been recorded from the nurseries and plantations of all five host plant systems. These include ten larval parasitoids, six pupal parasitoids and 12 species of insect predators. Along with insect predators, six species of spiders were also recorded from *Acacia nilotica* plantations. The natural enemy complexes were found in abundance, particularly in *Acacia nilotica* followed by *Tamarindus indica, Pongamia glabra, Azadirachta indica* and *Albizzia lebbeck.*

Social insects in Social forestry

Termites constitute the principal silvicultural pests in the tropics, where they cause economic loss in forest nurseries and during the establishment of young out-planted saplings or cuttings in reforested areas. Though well documented, reliable estimates in most cases, are lacking, they can be substantial particularly the exotics, which are susceptible to the native fauna.

Termites as pests of plantation forestry

All forestry crops are prone to be attacked by termites, particularly during the seedling stages in nurseries and young plantations, though the intensity and incidence of attack vary from species to species and from one locality to another. Species responsible for damage in plantation forestry, belong almost exclusive to family Macrotermitidae and the most wide spread damage is caused by some species of the genera *Macrotermes* (*M.estherae*) and *Odontotermes* (*O. distans, O. feae, O. obesus, O. microdetatus* and *O. indicus*). Localised attack by very few species (*O. horni, O. redemanni, Microtermes incetoides, M. mycophagus*) is common in some areas. The economically important wood decaying genus *Coptotermes* (*C. heimi*) may occassionally attack well grown and mature trees in plantations.

Fig. 1. A. Larva of *Euproctis lunata*. B. *Euproctis lunata* larvae feeding on the bark of Acacia nilotica. C. Defoliated tree of *Acacia nilotica* by *Euproctis lunata*.

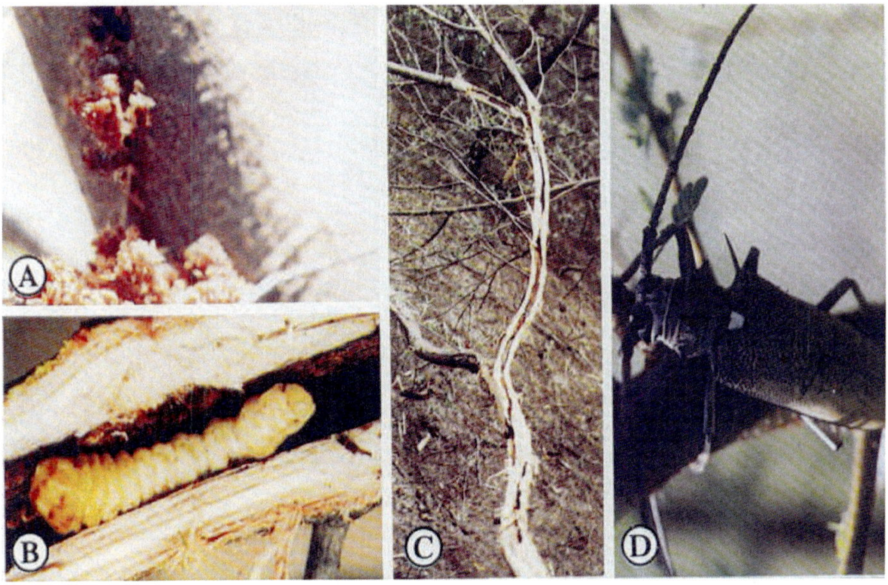

Fig. 2. A. Cankers on *Acacia nilotica* by longicorn beetle, *Coelosterna* sp. B. Larva of *Coelosterna* sp. C. Damaged tree of *Acacia nilotica* by *Coelosterna* sp. D. Adult *Coelosterna* sp.

Fig. 3. A. *Achaea janata,* B. Semilooper larva.

Eucalyptus, the most favoured and widely planted species in plantation forestry in India suffers severe damage from several species of subterranean and mound building termites in some parts of Andhra Pradesh, Maharashtra, Tamil Nadu and Uttar Pradesh. Similarly clones of some species of poplars (*P. deltoides, P. eumericana*) have been attacked by *Coptotermes heimi, Heterotermes indicola, O. distans,* and *O. obesus.* Young seedlings of *Acacia albeda,* and *A. tortilis* also suffer injuries from *Odontotermes microdentatus, O. obesus* and *Microtermes obesi* in forest nurseries in some parts of Haryana and Uttar Pradesh.

Incidence and nature of damage

Termite attack in plantation forestry may occur in three different stages, viz., (1) to seedlings or cuttings in forest nurseries, (2) to young plants of 1-3 years old in young plantations and (3) to mature trees in established plantations. In young seedling the attack generally takes place in the upper 20 cm of the soil layer. The termites feed on tap root as well as on rootlets, which results in the tapering or complete severance of the root system, leading ultimately to the death of the plant. The mortality is great during the first year (4-6 months in the nursery, 4-6 months after planting), when even 100% mortality can occur, though the damage during second year and subsequent years is not an uncommon phenomenon. The ecological characteristics of the land prior to the establishment of the crop and the extent to which it has been subjected to physical disturbances largely influence the termite attack. It is far greater on land, which has been previously under agricultural land use management, rather than on virgin land. The termite attack in plantation forestry may be (1) primary causal factor for mortality of healthy and vigourously growing seedlings in forest nurseries (4-6 months) and in young plantations, (2) Secondary causal factor for mortality in seedlings primarily killed by other disposing factors, notably drought and fungal pathogens and (3) complementary causal factor for mortality in seedlings including termites, where the actual cause is obscure, but which may be the result of combined actions of more than one factor.

Thakur (1990) proved that the control of *Microtermes* sp. in palm plantations is possible with simple irrigation methods. *Microtermes* sp. cannot have any activities in insufficiently irrigated plots and such irrigated plots are not suitable for its survival. Therefore, sufficient and regular irrigation of plots is the best way to control termites.

Table 1. Number of termite individuals in palm trees in experimental plots.

Replication/Treatment	1	2	3	4	Total	Average
Sufficiently irrigated plots	50	0	40	110	200	50
Insufficiently irrigated plots	2200	2750	2245	5300	13495	3373

(Courtesy: M. L. Thakur, 1990. Termites injurious to plantation forestry in India, 11th International Congress, IUSSI, India)

Lepidoptera

Lepidopterans are of great importance in larval stages. These are phytophagous and act as serious pests, defoliators, pollinators and parasites. Majority of species devour foliage and shoots of trees and crops. A smaller number bore into stems or attack underground parts. Its presence or absence serves to monitor ecological changes in the particular habitat.

Pteroma plagiophleps is a defoliator of *Tamarindus indica* in India since 1970s. In recent years it caused extensive defoliation in young plantations of *Albizia falcateria* in Kerala. It has become an important serious pest in South and Southeast Asia. The insect appears to be gradually extending their host range and are important as a forest pest attacking plantations of *Acacia nilotica* and mangroves, *Rhizophora mucronata*. A 28-36% of population eradication of this pest has been noticed due to Ichneumonid and Chalcidid wasps. There are large numbers of *Acrocercops* spp. which mines the leaves of wasteland forest tree species *Syzygium cumini*. *A. telestris* is a pest of cinnamon. *A.gemoniella* mines the leaves of nursery stocks of milk tree, *Manilkara hexandra* damaging upto 20% of leaves in some nurseries of central India. The bark eating caterpillar *Indarbela quadrinotata* is a polyphagus pest infesting many fruit trees, and forest trees such as *Tectona grandis, Sweitienia* spp., *Gmelina arborea*, and species of *Acacia, Terminalia, Casuarina, Syzygium* and *Emblica* etc in India. Broad irregular patches are excavated in the bark and these areas are roofed in, mixed with silk and excrement and fragments of bark. Feeding by this caterpillar can result in the loss of growth increment and with heavy infestation, gridling and mortality. This pest has been reported as a serious pest of *Acacia senegal* and *A. tortilis* in Thar desert. These tree species are major component of plantation forestry programmes related to wasteland development programme, sand dune stabilization, fuelwood and fodder for this arid region. Natural enemies of this pest include braconid and eulophid wasps and ant predators.

Coleoptera

Celosterna scabrator, commonly known as babul borer because it attacks Acacia species such as *A. arabica* and *A. nilotica*. It is also a pest of teak, *Casuarina*, *Prosopis* and *Dipterocarpus*. Feeding by larvae causes cessation of growth and some times death of the tree. Incidence of attack of up to 80% has been recorded where the trees are growing on unsuitable sites. The pest attack on *Eucalyptus* spp. plantations is common in India. It is believed to be responsible for up to 14% mortality of *Prosopis cineraria*, an important agroforestry tree species in North India. In Karnataka outbreaks have occurred in one year old plantations of *E. tereticornis* and *A. nilotica* with high mortalities. Feeding by adults can result in girdling and breakage of the main stem and branches. Few natural enemies have been recorded. *Oxyrachis tarandus* is reported to be the most notorious pest of *A. nilotica* in various localities.

Calopepla leayana is the most destructive defoliator of *Gmelina arborea*. It is known to destroy more than 2000 acres of young plantations in a single year in India. It has been reported that trees remain leafless for almost half the period of the growing season due to defoliation by this pest. In Northeastern India, monoculture of *G. arborea* has been abandoned in the past because of this insect. Singh *et al.* (2006) reported an exhaustive account on the bioecology of a promising biocontrol agent namely *Brachymeria excarinata* (Hymenoptera: Chalcidae) and found that wasp might have an immense biocontrol potential in managing the dreaded beetle. Entomopathogenic fungi viz., *Beauveria bassiana* and *Metarrhizium anisopliae* were found to be significantly pathogenic to all the larval as well as adult stages of *C. leayana*. The fungi cause white and green muscardine diseases to the larvae and adults respectively (Senthilkumar, 2005). *Acanthophorous seraticornis* a cerambiciid grub which is known to occur near the roots of mature teak trees is known to cause damage to roots resulting in the death of the tree. A good number of larvae have been observed near the roots.

Hemiptera

The lace bug, *Tingis beesoni* is a serious pest of *Gmelina arborea* in India is causing defoliation and die back in young plantations. Feeding by nymphs and adults causes leaves to become blotched brown near the base, they wither and fall and the shoots become dry. Eventually shoots die back and retard the tree growth. During one outbreak in a 10 ha plantation of 1 year old trees in India, 67% of the plants were infested and defoliated and die back of the terminal shoot. An incidence of the teak scale insect *Icerya aegyptiaca*, was found on plantations of *Tectona grandis* in Northeast (Senthilkuamr and Barthakur, 2005). Out of 50 clones, 9 clones were not infested with teak scale insect, followed by 27 clones with less infestation, 8 clones with moderate infestation. A total of 6 clones were heavily infested with the scale insect. An outbreak of the tea mosquito bug, *Helopeltis antonii* has been noticed in *Azadirachta indica* in Andhra Pradesh last year. It sucks the juice from the leaves and stem resulting in drying up of shoots. Affected trees exhibit burnt appearance. A few entomopathogenic fungi were found to be effective against this pest. The mango pest mealy bug *Drosicha stebbingi*, primary pests of *Mangifera indica* also attack *A. nilotica*. Heavy infestation caused by the bug results in the death of shoots and twigs and deforms the growth of the saplings of *A. nilotica*.

Hymenoptera

An invasive species *Leptocybe invasa* supposed to be a native of Australia was first recorded on Eucalyptus from Tamil Nadu in 2002. It is distributed in various parts of Andhra Pradesh, Pondichery, Kerala, Karnataka, Punjab, Uttaranchal, and has been noticed invariably in all the clones of Eucalyptus and more than ten species of Eucalyptus too. Wasps insert their eggs inside the newly developed leaves and the gall development takes place at five different stages right from oviposition site followed by typical bump shaped gall to gall with adult's emergence holes. It is a serious problem on *Eucalyptus* in India. Two species of Tetrastichinae, *Quadrastichus mendeli* and *Selitrichodes kryceri* (Hymenoptera: Eulophidae) from Australia are described as parasitoids of *Leptocybe invasa*. They are still under quarantine in India, to be tested under classical biological control programme. *Trichilogaster acaciaelongifoliae* is a serious gall inducing wasp on *Acacia longifolia,* reduced seed production by 95-99%.

Diptera

Saproxylic organisms are those dependent at some stage during their life cycle, on dead wood or decaying material associated with the woody parts of trees. Saproxylic organisms vary from fungi to woodpeckers but the most biodiverse groups are Coleoptera and Diptera. In Europe saproxylic organisms often require conservation because they are under threat due to severe reduction of natural woodland and impoverishment of what remains. A particularly important expression of concern is the recommendation to the countries to protect saproxylic organisms and their biotopes. However, in comparison with Coleoptera, saproxylic Diptera are poorly known. In contrast to Coleoptera, the majority of saproxylic Diptera need measures to conserve wet microhabitats for breeding. Coleoptera frequently require veteran (large, old) trees. Although veteran trees often have more wet microhabitats than young trees, they are not essential for saproxylic Diptera. Red-listed saproxylic Diptera were found breeding in small, young trees (Rotheray *et al.*, 2001).

Orthoptera

Trees planted are particularly susceptible to grasshopper damage because the grasshoppers prefer the leaves of many trees to grasses. When numbers of grasshoppers are high, mature trees can be completely defoliated. In addition to being planted in strips such as shelter belts and usually surrounded by crop, there is no understorey which may act both as a barrier to wingless grasshoppers and a refuge for their parasites. The common orthopterous insects that cause damage to forest, shade trees and trees in shelterbelts are acridids and tettigoniids. A few of the orthopteran insects are locally very common and can occassionally cause significant damage in localised areas. The grasshopper *Valanga nigricornis* caused sporadic defoliation in teak plantation in Sri Lanka. Nair (2007) reported eighteen species of orthopteran insect pests attacking teak in India. *Zonocerus elegans*, and *Gryllus bimaculatus* are very common orthopteran insect pests on commercial plantations of Eucalyptus and Acacia in South Africa.

PARASITOIDS IN SOCIAL FOREST STANDS

Most of the parasitoids were found affecting different development stages of lepidopteran pests which cause extensive defoliation of *A. nilotica, T. indica and P. glabra.* The tachinid, larvae of a polyphagous parasitoid, was found to play a major role in bringing down the populations of various pests such as *Euproctis lunata, Eumeta cramerii, Pteroma plagiophleps, Manatha assamica, Metanastria hyrtaca* on *Acacia nilotica* and *Tamarindus indica,* and *Dasychira mendosa, Selepa celtis* and *Tephrina pulinda* on *A. nilotica.* Apart from the tachinid, *Sinophorus psycheae, Brachymeria carinata,* and *Brachymeria marmonti* were encountered on *Eumeta cramerii, Pteroma plagiophleps* and *Manatha assamica.* While these three are confined to bagworms, *S. psycheae* is a larval parasitoid, and *Brachymeria* spp. is a pupal parasitoid. *Acanthrmius* sp. and *Goryphus* sp. were specific to *P. plagiophleps. Goryphus* sp. is a pupal parasitoid, preferring the female pupa only. Species such as *Apanteles cirphicola, Tetrastichus krishnieri, Charops* sp. and *Eurytoma braconidis* were also obtained from *E. lunata* infesting *A. nilotica. Apanteles cirphicola* and *Enicospilus* sp. were the other two larval parasitoids found in association with *T. pulinda.* Species such as *Pseudagenia* sp., *Brachymeria circulae* and *Microdus fumipennis* were found in association with *M. hyrttaca* which caused defoliation in both *A. nilotica* and *T.*

indica. Of these, *M. fumipennis* occurred in *A. nilotica* while other two were found in association with *T.indica* and *A. nilotica*. The bagworm *Thyridopteryx ephemeraeformis* feeding on *A. nilotica, T. indica* and *P. glabra* was parasitized by a braconid *Phaneratoma* sp. *Kriechbaumerella ayyari* is a pupal parasitoid of *Latoia lepida* attacking *T. indica*. These parasitoids were found in abundance at the time of pest outbreaks of their respective hosts. *Apanteles cirphicola* is a polyphagous parasitoid, parasitizing *E. lunata* and *T. pulinda* in *A. nilotica*.

Outbreak of pest populations in nurseries and plantations of social forest stands is checked by the parasitoids (Tables 2 and 3). Significant correlation between the fluctuation of the pests and their natural enemies was observed based on parallel changes in their relative population size and the host population determines the population build-up of parasitoids. For example, *Euproctis lunata* a serious defoliator of *A. nilotica*, and occasionally on *T. indica* attained a very high population in April and began to decline during the succeeding months. However, higher population levels were attained during October and February. A significant parallel increase in parasitoid populations was evident with their host-insect population build up leading to a natural parasitism of upto 40 per cent in the field. Various other parasitioids constituted a total of 22 per cent parasitism. The tachinid parsite played a major role in checking *E. lunata* population which constitutes upto 31 per cent, followed by Charops sp., *Tetrastichus krishnieri* and *Eurytoma braconidis*. *Tephrina pulinda* was associated with the nurseries and plantations of *A. nilotica*. Its population showed a bimodal peak during September and February, with maximum incidence in September. *A. cirphicola* and the tachinid parasitoids affected a total of 13.0 per cent parasitism, of which, *A. cirphicola* was upto 8 per cent and tachinid 5 per cent. Another parasitoid, *Enicospilus* sp., was noticed at the time of pest outbreak.

PREDATORS IN SOCIAL FOREST STANDS

Among 12 predators, reduviids constituted the major predatory group in *A. nilotica*. The reduviids such as *Irantha armipes, Raphidosoma atkinsoni, Acanthaspis siva, Brassivola fuscipes* were found to be active predators against the immature stages of *T. pulinda, E. lunata, E. cramerii, S. celtis, D. mendosa on A. nilotica. A. siva* was found restricted to *T. indica* plantation where it fed on *Myllocerus* spp. *Rhinocoris fuscipes* was to *T. indica* plantation where it fed on *Parnara mathias* infesting *P.*

glabra. Euproctis lunata showed bimodal peak during March and January. In September its population was very low (20 per plant) from April onwards the population exhibited a gradual decrease followed by population build-up during October and the following months. Effective suppression of the population build-up of the bagworm *E. cramerri* on *A. nilotica* by various species of parasitoids, noted in this study corroborates the earlier accounts incorporating such positive information on this subject by several other workers.

Table 2. Relationship between *Acacia nilotica* and *Tamarindus indica* Herbivore Insect and Natural Enemies

Acacia nilotica		***Tamarindus indica***	
HOST INSECTS	**PARASITOIDS**	**HOST INSECTS**	**PARASITOIDS**
Eumeta cramerii	*Sinophorus psycheae*	*Chaliodes ferecitrea* *Manatha assamica*	*Sinophorus psycheae*
Pteroma plagiophleps *Thyriopteryx*	*Acanthormius* sp *Phanerotoma* sp	*Pteroma plagiophleps*	*Acanthormius sp*
E. cramerii *C. ferevitrea* *M. assamica* *P. plagiophleps* *Euproctis lunata* *Selepa celtis* *Dasychira mendosa*	*Tachinid*	*E. cramerii* *C. ferevitrea* *M. assamica* *Metanastria hyrtaca* *P. plagiophleps*	*Tachinid*
E. lunata	*Charops* sp. *Tetrastichus krishnieri* *Eurytoma braconidis*	*P. plagiophleps* *M. assamica* *E. cramerii*	*Brachymeria carinata* *Brachymeria marmonti*
T. pulinda *E. lunata*	*Apanteles cirphicola*	*P. plagiophieps*	*Goryphus* sp.
T. pulinda	*Enicospilus* sp.	*M. hyrtaca*	*Brachymeria criculae* *Pseudagenia* sp. *Microdus fumipennis*
E.cramerii *P. plagiophleps*	*Brachymeria carinata* *Brachymeria marmonti*	*Latoia lepida*	*Kreichbaumerella ayyari*
P. plagiophieps	*Goryphus* sp.		
M. hyrtaca	*Brachymeria criculae* *Pseudagenia* sp. *Microdus fumipennis*		

Table 3. Interrelationship between *Acacia nilotica* and *Tamarindus indica*, Herbivore Insects and Natural Enemies

HOST INSECTS	PARASITOIDS
Acacia nilotica	
Tephrina pulinda	*Eocanthecona furcellata*
Eumeta cramerii	
T. pulinda	*Rhinocoris fuscipes*
	Irantha armipes
Euproctis lunata	*Brassivola* sp
E. cramerii	
Coccids	*Chrysopa scelestes*
T. pulinda	*Peucetia viridana*
	Oxyopes rufisternum
	Stegotyphus sp.
	Argiopes sp.
	Crab spider
	Plexippus sp
Weevils	*Oxyopes rufisternum*
Grasshoppers	*Stegotyphus* sp.
	Argiopes sp.
	Crab spider
	Plexippus sp
Tamarindus indica	
Latoia lepida	*Eocanthecona furcellata*
Eumeta cramerii	
E. cramerii	*Plexippus* sp

Parasitoid and predator complexes in social forest stands are important components to reduce pest attack in nurseries and plantations. These natural enemy complexes play a major role in controlling the outbreaks of insect pests such as *Eumeta cramerii, Pteroma plagiophleps, Tephrina pulinda, Dasichira mendosa* on *Acacia nilotica*, of *Pteroma plagiophleps* and *Manatha assamica*, on *Tamarindus indica*; and of *Parnara mathias* on *Pongamia glabra*. Field observations on the parasitization rate of different parasitoids like the techinids (Diptera), *Sinophorus psycheae* (Ichneumonidae), *Brachymeria carinata* and *Brachymeria marmonti* (Chalcidae), reveal 65-70 percent parasitism on bagworms.

Among these parasitoids, tachinids showed a higher percentage of parasitism on *E. cramerjj* followed by *S. psycheae*. Comparative field parasitization rates of the parasitoids, viz., *Apanteles cirphicola* (Braconidae), *Phanerotoma* sp. (Braconidae), tachinid and *Charops* sp. (Ichneumonidae) on *E. lunata* and *T. pulinda*, indicate that *A. cirphicola* is an efficient biocontrol agent compared to other parasitoids. Among the predatory resources, the reduviid *Rhinocoris fuscipes* a polyphagous predatory stink bug *Eocanthecona furcellata*, are potential biological control agents.

Ananthakrishnan (1996) suggested that structure, composition and the chemical and physical properties of the litter are important for the species assemblage of litter dwelling communities, which agrees with the low resilience of these communities observed by other authors. Therefore, in the case of short intervals between fires (7-10 yr. in our study at sites repeatedly burnt), the litter hardly recovers and conditions remain unstable. On the other hand, we observed a positive effect on epigaeic and mobile groups, in particular with Araneae and Carabidae. Many authors suggest that these groups, after having been reduced during the fire, increase quickly by profiting from abundant food in the post-fire mosaic of ground habitats. The post-fire conditions of herbs, shrubs and trees seem to favour heliophilous and floricol insects (e.g. Aculeata and Syrphidae). Ananthakrishnan (1996) emphasized that the growing awareness of the significance of biodiversity and its rapid decline, has led to increasing interest in inventorying and monitoring of biodiversity at various levels alongside priorities for their conservation. He also stressed that the general assessment of biodiversity at the genetic, population, specific and ecosystem levels in vogue, increasing involvement in insect biodiversity studies has to necessarily involve multidimensional aspects such as forest canopy, understorey and forest floor encompassing 'species guilds' in tropical forest, besides agro biodiversity involving genetic diversity found within individual crop and associated weeds..

Future priorities

As a result of the review discussed above, there is a new impetus behind tree protection in industrial and other plantations. Overall, given that there is a continuum of land use systems involving trees in India, from pure monocultures through to mixtures of crops (including agricultural crops), pest management strategies need to have an integrated approach, i.e., utilize integrated pest management (IPM), which includes sustainable, low technology, environmentally sound and cost-effective components appropriate to particular land use systems and local conditions. Important components that should be considered in an integrated approach include: pest monitoring schemes, biological control, which includes the conservation, augmentation, or introduction of natural enemies, host plant resistance, and silvicultural practices. Particular emphasis needs to be given to IPM strategies that prevent or reduce the risk of pest problems occurring. In this context, particular attention should be paid to practices such as correct site matching, conservation of natural enemies, etc. These factors must be taken into consideration before planting programmes begin. Further to this, the success of any IPM strategy depends on its careful implementation; thus IPM programmes should contain a significant training component for forest monitors and local farmers.

Conclusion

Globally the forests are in a state of transition: from natural to managed forests, with changing biodiversity, from an era of over-abundance with little concern for the environment, to a period of tighter wood supplies, with increasing consciousness for the environment and the ecosystem. Forestry, in general, has progressed from the concept of harvesting and regeneration to an integrated management of the forest resources and ecosystems. Along with this, protection of forests from pests and other disturbances has also changed from control and eradication to more effective longterm management of pests within the concept of ecosystem. Current trends in forest health point away from the antagonistic approach of attempting to eradicate pests and towards working with natural processes in the context of ecosystem management. To be successful, this approach does not require a perfect understanding of ecosystems – only explicit statements of our assumptions, our knowledge of the processes and interactions, and an understanding of the possible actions to be taken and their consequences. To assess the potential harm of a particular disturbance to the forest ecosystem, we must develop better understanding of the initial conditions of the forest. This will indicate if the disturbance is operating outside the historic range, in terms of intensity and frequency and will allow us to determine its impact and what compensatory actions should be taken. The philosophy of ecosystem management, incorporating the adaptive potential of experimentation and research, can direct our actions to maximize our use of natural systems and minimize any harmful impacts of crisis management strategies, by placing us back into the context of our natural world. Short-term tactics must be employed with long-term goals and strategies, applied on a site-specific basis which recognizes the ecological capabilities of each site.

Almost all the major forest insect pests of India are indigenous to this region and have an associated complex of natural enemies (predators, parasites and pathogens) that have evolved with them, with fluctuating population levels, constantly interacting with one another. When certain changes occur in the ecosystem, such as the introduction of monocultures reducing or eliminating plant diversity, excess or lack of rainfall and drastic temperature differences, the natural biotic regulating mechanism may break down and lead to pest outbreaks. For agumentative or inundative releases of native parasites or predators the selection of the best suited and most promising species is very important. Ready amenability to mass multiplication on the natural host or on unnatural hosts in the insectary should never be the sole or over-riding criterion because the control value of a natural enemy is to be seen in the field with all its vicissitudes over a period of time after release. Therefore the selection of species of a natural enemy for mass-production and releases should be based on a close study of the diverse aspects of their biology, phenology, behaviour etc., and on a proper evaluation of all species attacking different host stages at various host densities in normal and outbreak situations. Some parasites have inherently disadvantageous biological parasites traits. For example, *Rhyssa persuasoria* is reported to be an inefficient parasites or Siricid wood-wasps, because of its poor host-searching and oviposition ability. The female spends a lot of time and energy boring into areas of tree trunks with no host larvae, it produces only 24 mature eggs and takes ten or more minutes for each successful oviposition. A polyphagous parasite like *Cedria paradoxa* with several hosts on a wide range of plant species would be a poor choice for massbreeding and release.

There are some outstanding cases of successful biological control of native insects by parasites of the same or closely related pest species or genera from other geographical areas. *Lymantria obfuscata* and *Hypsipyla robusta* are two major forest pests for which parasites of allied species from the neotropics are worth importing and for field trials. For some of the other major forest pests, from their native distribution range and also from the occurrence of allied species with similar biological features and ecological parameters in other parts of the world potentially valuable biotic agents may be selected for evaluation in India. The only instance of classical biological control of a forest insect pest in this country is the recent introduction of a readily available beetle predator, *Curinus coeruleus* with its limitations may eventually turn out to be a general predator like *Cryptolaemus montrouzieri*. It is polyphagous and was introduced from Mexico into Hawaii as far ago as in 1922 as one of the three species of coccinellied predators to control the mealybug *Nipaecoccus nipae* on coconut palms. Although it soon became established in Hawaii, complete control of the mealybug was achieved only by the introduction of a host-specific hymenopterous parasite. Similar neotropical parasites of the *Leucaena* are currently being investigated elsewhere in Asia and these should be introduced here without delay.

As discussed earlier, native natural enemies of many forest pests are normally able to keep the host populations below the economic injury level but since they are themselves subject to attack by their natural enemies and to various other biological and environmental stresses, and they become ineffective, it is helpful to augment their populations in the field by breeding them in insectaries and making one or more releases at the appropriate time. If the releases have to be repeated periodically the total cost of production, transportation and field release has to be considered in relation to the actual benefits. Forest areas are largely managed by government organisations which often lack the required financial, material and manpower resources to undertake such biological control campaigns over extensive forest areas. Therefore, the introduction of more efficient biocontrol organisms from other geographical but ecologically similar regions is very worthwhile and this alternative approach merits serious consideration. If one or more of the introduced natural enemies provide adequate and sustainable biological control there is no need for recurrent efforts and expenditure.

References

Ananthakrishnan, T.N. (1996). *Forest litter insect communities,* Science Publishers, Lebanon.

Beeson, C.F.C. (1941). The ecology and control of the forest insects of India and neighbouring countries. 1993 reprint Bishen Singh Mahinderpal Singh, Dehradun, India.

Borota Jan (1991). Tropical forests : some African and Asian case studies of composition and structure, Elsevier Science Pub. Co, Canada. xii, 274 p.

Chaterjee, P.N. and Sen-Sarma, P.K. (1962). Seasonal incidence of wood destroying subteranean termites. *Indian Forester*, 88, 139-141.

Chatterjee, P.N. and R.S. Thapa 1964. Recent epidemic of sal borer *Hoplocerambyx spinicornis* in South Mandla Forest Division (Madhya Pradesh) and recommendation for control. *Indian Forester*. **90** (11): 777-781.

Duelli P, Obrist M.K. and Flückiger P.F. (2002). Forest Edges are Biodiversity Hotspots – Also for Neuroptera, Acta Zoologica Academiae Scientiarum Hungaricae **48(2):** 75–87.

Hunter, M.D. (2002), Landscape structure, habitat fragmentation and the ecology of insects. *Agricultural and forest entomology*, 4, 159-166.

Mathur, R.N. and Singh, B. (1960). A list of forest insect pests of India and the adjacent countries. *Indian Forest Bulletin*, 171 (9).

Moretti M., Obrist M.K. and Duelli P (2004). Arthropod biodiversity after forest fires: Winners and losers in the winter fire regime of the Southern Alps, *Ecography* **27:** 173-186.

Nair, K.S.S. (2007). Tropical insect pests: Ecology, impact and management, Cambridge University press, London.

Rotheray, G.E., Hancock, G., Hewitt, S., Horsfield, D., MacGowan, I., Robertson, D. and Watt, K. (2001). The biodiversity and conservation of saproxylic Diptera in Scotland, *Journal of Insect Conservation*, **5:** 77–85.

Scherber, C., Milcu, A., Partsch, S., Scheu, S. and Weisser, W.W. (2006). The effects of Plant Diversity and Insect Herbivory on Performance of individual Plant Species in Experimental Grassland, *Journal of Ecology, doi*: 10.1111/j.1365-2745.2006.01144.x

Senthilkumar, N and Barthakur, N.D. (2005). A record of Egyptian fulted scale, *Icerya aegyptiaca* on teak in Assam. *Insect Environment,* 11(1): 31-32.

Senthilkumar, N. (2005). Effect of entomopathogenic fungi on different stages of *Calopepla leayana* Latr. (Chrysomelidae : Coleoptera), a major defoliator of *Gmelina arborea* (Roxb.) (Verbinaceae) *Insect Environment*, 11(3):108-111.

Singh, S., Kumar, A., Senthilkumar, N. and Singh, A.N. (2006). Strategies for the management of *Craspedonta leayana* (Coleoptera:Chrysomelidae) in *Gmelina arborea*. *Indian Forester* 132 (5):581-588

Stork N.E. and Grimbacher P.S. (2006). Beetle assemblages from an Australian tropical rainforest show that the canopy and the ground strata contribute equally to biodiversity, Proc. R. Soc. B **273:** 1969–1975.

Thakur, M.L. (1990). Termites injurious to plantation forestry in India, 11th International Congress, IUSSI, India.

Chapter - 8

Insect Biodiversity in Scrub Jungle Environment

R.W. Alexander Jesudasan

Department of Zoology, Madras Christian College (Autonomous), Tambaram East, Chennai 600 059

Introduction

Scrub jungles are characterized by plants with adaptations to dry climate, with features to overcome water loss, succulent leaves and stems and storage organs to store water. While tropical forests are structurally more complex along the habitats to accommodate more niches, scrub jungles with limited vegetation have limited components. While insect feeding guilds are limited, though noticeable, the severity of their injury is often limited with such common inhabitants like the foliivores, namely caterpillars and sap feeders like aphids, coccids, whiteflies, etc. It must also be realized that plants that form extensive species-poor communities tend to have different defensive strategies than those of species-rich communities, living as they do in highly stressful environments; many species evolve better defenses for lowering insect damage. Nevertheless, the diversity of species in scrub jungles combined with the vagaries of climate, has an important bearing on species diversity. Ecological conditions can exert a powerful influence on insect-plant interactions and under conditions of stress. The diversity of inter specific competition prevails, with a correlation between seasonal development and food plants available.

Observations presented here pertain to the scrub jungle vegetation prevailing in the Madras Christian College campus extending to an area of about 365 acres, a unique eco-system with notable floral and faunal composition, and also the second largest scrub jungle in Asia.

Plant diversity include about 458 plants belonging to 90 families of which 115 are trees, 66 are shrubs, 56 are climbers,

202 are herbs and 19 are stragglers. Besides the native plants, a large number of these plants are exotic but have now become a part of the ecosystem. These include Angiosperms, Gymnosperms, Pteridophytes and Bryophytes. Angiosperms, the flowering plant group, depict the dominant plant diversity on this large campus. They show diversity in terms of their height, nature of habitat and economic value viz., edible, medicinal, ornamental, timber-yielding, fibre-yielding, dye-yielding, etc. Some of the common trees are *Albizia lebbeck, Diospyros chloroxylon, Lannea coromandelica, Syzygium cumini, Albizia amara, Peltophorum pterocarpum, Tebebuia rosea, Tamarindus indica, Cassia sp., Delonia regia.* There are also plants like *Atalantia monophylla, Bauhinia racemosa, Drypetes sepiara, Grewia orbiculate, Capparis brevispina, Combretum albidum, Hugonia mystax, Ventilago madraspatana* and *Ixora pavetta.*

Insect Diversity in Scrub Jungles

While homopterans abound, heteropterans are limited, with sufficiently large number of thrips species, Lepidoptera and Hymenoptera, with restricted number of Dictyoptera, phasmids, Orthoptera, Dermaptera, Isoptera, Neuroptera and Coleoptera. Among the Homoptera, whiteflies or aleyrodids dominate with sizable number of cercopids or spittlebugs, jassids, leaf hoppers and aphids. Thrips or Thysanoptera are predominant as also aphids and Lepidoptera. Some of the more important species belonging to the above groups are indicated below:

A List of Scrub Entomofaunal groups collected in the scrub jungle of Tambaram

Order: *Hemiptera*

Homoptera

Whiteflies (list appended)

Membracids

Spittle Bugs

Jassids

Aphids - *Myzus persicae*

Heteroptera

Halys sp.

Spilosthethus sp.

Iphita limbata

Stink bugs - *Aspongopus janus*

Nezara viridula

Order: *Thysanoptera*

Thrips

Order : *Dictyoptera*

Cockroaches – *Theria petiveriana, Blatta* sp.

Praying mantids

Order : *Phasmida*

Walking sticks, Leaf insects

Order : *Orthoptera*

Grasshoppers, Katydids

Order : *Dermaptera*

Earwigs – *Euborellia annulipes*

Order : *Hymenoptera*

Paper wasps, Bumble bees, ants – *Oecophylla* sp. (nest-building ant)

Order : *Lepidoptera*

Butterflies - *Papilio demoleus, Euploea core, Danaus plexippus, D. chrysippus*

Moths – Bag-worm moths

Order : *Isoptera*

Termites

Order : *Neuroptera*

Antlions

Order : *Coleoptera*

Fire flies, Dung rollers, Bark beetles

Whitefly interaction with ants and other insects

Whiteflies as a group have gained increasing importance over the years as one of the most noxious insects because of their tendency to inflict significant damage to agricultural and horticultural crops through their numerical strength. A large number of species of whiteflies are polyphagous in nature and have the uncanny ability to occupy a wide range of host plants worldwide, though they are found mostly in the tropical, subtropical and temperate regions. What they lack in size they make up by their numbers and by their aggregate feeding, causing severe damage to young leaves, blossoms and fruits leading to premature dehisence, defoliation and in turn affecting the entire plant. They are also known to transmit viral diseases in plants. Till recently, whiteflies were considered as pests of minor importance but they have now reached the status of serious pests infesting vegetable crops and economically important trees like *Tectona grandis, Dalbergia lanceolaria, Dalbergia sisoo* and *Syzygium cumini*, and several reasons have been ascribed to their recent severe outbreak.

An interesting association of whiteflies with other insects like psyllids, scale insects, leaf hoppers and arachnids like spiders and mites has been observed. The excrement of whiteflies called honeydew is a major part of the diet of various insects such as wasps, different species of flies and ants. The whitefly-ant interaction is an interesting symbiotic association exhibiting trophobiosis, and provides for the successful survival of both the organisms. Whiteflies benefit from ant tending because the presence of ants prevents other insect predators from attacking the host plants and prevents the build-up of fungus that develops on uncollected honeydew.

The whiteflies, *Rhachisphora indicus* and *Rhachisphora trilobitoides* were found to infest 5 host plants namely *Ixora pavetta, Memecylon edule, Memecylon umbellatum, Syzygium cumini* and *Benkara malabarica* attracted a large number of diverse species of ants.

The exudation of honeydew by the nymphal instars of this whitefly attracted 7 different species of ants: *Solenopsis germinata, Camponotus* sp., *Meranoplus bicolor, Myrmicaria brunnea, Paratrachina longicornis, Tapinoma melanocephalon* and *Crematogaster* sp. The pattern of activity differed in each of the ant species as each species had its own time period for foraging activity which was dependent on the temperature and on the intensity of light, rainfall and weather.

Some of the common whiteflies occurring in Tambaram are

- *Acaudaleyrodes rachipora*

- *Aleurocanthus woglumi*
- *Aleuroclava complex*
- *Aleurodicus dispersus*
- *Aleurolobus marlatti*
- *Aleurolobus orientalis*
- *Aleuromarginatus kallarensis*
- *Aleuroplatus alcocki*
- *Bemisia tabaci*
- *Dialeuropora murrayae*
- *Rhachisphora trilobitoides*

Membracids – Tree hoppers or Cow bugs

Treehoppers (more precisely *typical treehoppers* to disting-uish them from the Aetalionidae) and *thorn bugs* are members of the family *Membracidae,* a group of insects related to the cicadas and the leafhoppers. There are about 3,200 species of tree-hoppers in over 600 genera. They are found on all continents except Antarctica, although there are only three species in Europe.

They are best known for their enlarged and ornate pronotum, which most often resembles thorns, apparently to aid camouflage. But in some species, the pronotum grows to a horn-like extension, and even more bizarre and hard-to-describe shapes are also found. Thorn bugs, due to their unusual appearance, have long interested naturalists.

They suck the sap from the trees, shrubs and cultivated crops and most species are host specific. Principal damage is caused by egg laying. Ants are found associated with the membracids. Ants stroke the bugs by their antennae which makes them exude honeydew, which the ants ingest. *Telengana consabrina* is a cow bug found on MCC campus.

Spittle Bugs

They belong to Family Cercopidae and are also known as froghoppers or cuckoo-spits. They are small, jumping insects usually brown or grey or sometimes variously coloured. They feed on shrubs and herbaceous plants. The nymphs always remain inside a frothy spittle mass resting head downward on the plant and hence the name 'spittle bugs' and this is a common sight in meadows and on grasses along rice fields. In each frothy mass one of more brownish or greenish nymphs can be seen. They leave the spittle after the last moult and become active. The fluid voided from the anus and the mucilagenous secretion from the epidermal glands on the seventh and eighth abdominal segments form the frothy mass. The caudal appendages aid in introducing air bubbles into the spittle. Spittle is not produced by the adult insect.

Jassids

Leafhopper is a common name applied to any species from the family *Cicadellidae.* Leafhoppers, colloquially known as *"hoppers"*, are minute plant-feeding insects in the superfamily Membracoidea in the order Hemiptera. Leafhoppers are found all over the world and constitute is the second-largest family in the Hemiptera.

Leafhoppers have piercing-sucking mouthparts, enabling them to feed on plant sap. A leafhoppers' diet commonly consists of sap from a wide and diverse range of plants, but some are more host-specific. Leafhoppers mainly are herbivores but some are known to eat smaller insects such as aphids on occasion. A few species are known to be mud-puddling, but as it seems females rarely engage in such behavior. Leafhoppers can transmit plant pathogens such as viruses, phytoplasmas and bacteria. Cicadellidae species that are significant agricultural pests include the Green Rice Leafhopper (*Nephotettix nigropictus*), Mangohopper and White Rice leafhopper (*Tettigella spectra*).

In some cases the plant pathogens distributed by leafhoppers are also pathogens of the insect themselves and can replicate within the leafhoppers' salivary glands. Leafhoppers are also susceptible to various insect pathogens, including Dicistroviridae viruses, bacteria and fungi; numerous parasitoids attack the eggs and the adults provide food for small insectivores.

Aphids

Aphids, also known as *plant lice* are small plant-eating insects, and members of the superfamily *Aphidoidea.* Aphids are among the most destructive insect pests on cultivated plants in many temperate and tropical regions. The damage they do to plants has made them enemies of farmers and gardeners the world over, but from a purely zoological standpoint they are a very successful group of animals.

About 4,400 species of 10 families are known. Historically, many fewer families were recognized, as most species were included in the family Aphididae. Around 250 species are serious pests for agriculture and forestry as well as an annoyance for gardeners. They vary in length from one to ten millimetres.

Natural enemies include predatory lady beetles (Coleoptera: Coccinellidae), hoverfly larvae (Diptera: Syrphidae), parasitic wasps, aphid midge larvae, crab spiders, lacewings (Neuroptera: Chrysopidae), and entomopathogenic fungi like *Lecanicillium lecanii* and the Entomophthorales.

Myzus persicae is a small green aphid. It is the most significant aphid pest of peaches, causing shrivelling of the leaves and necrosis of various tissues. Among the natural enemies of the *M. persicae* are both predators and parasitoids, including: beetles such as the ladybirds *Adalia bipunctata*, *Coccinella septempunctata* and *Coccinella decempunctata*; true bugs such as the anthocorids *Orius* and *Anthocoris*; neuropterans such as the chrysopids *Chrysopa* and *Chrysoperla*; syrphid flies such as *Syrphus, Scaeva, Episyrphus*; cecidomyid flies such as *Aphidolestes*; and aphids such as *Aphidus*.

Mutualism in Aphids

Aphids share a mutualistic relationship with some species of ants. The ants feed on the honeydew excreted by the aphids and in turn protect the aphids from predators. Some species of ants "farm" aphids, protecting them on the plants they eat, eating the honeydew that the aphids release from the terminations of their alimentary canals. This is a "mutualistic relationship". These "dairying ants" "milk" the aphids by stroking them with their antennae. Therefore, sometimes aphids are called "ant cows".

Some farming ant species gather and store the aphid eggs in their nests over the winter. In the spring, the ants carry the newly-hatched aphids back to the plants. Some species of dairying ants manage large "herds" of aphids that feed on roots of plants in the ant colony. Queens that are leaving to start a new colony take an aphid egg to found a new herd of underground aphids in the new colony. These farming ants protect the aphids by fighting off aphid predators.

Red bug, *Iphita limbata* (Stal.)

This phytophagous bug belongs to the family Pyrrhocoridae, which include brightly coloured (red, being predominant) and conspicuously large (ranging from 0.25 to 2 inches) polyphagous bugs, enjoying a wide host range. This insect is distributed in abundance in the scrub jungles of the MCC campus, predominantly under the trees of *Sterculia foetida* (Sterculiaceae) where they feed mainly on the fallen seeds.

Stink Bugs

Pentatomoidea is a superfamily of insects among the Heteroptera and are commonly referred to as *shield bugs* or *stink bugs.* Shield bugs have glands in their thorax between the first and second pair of legs which produce a foul smelling liquid. This liquid is used defensively to deter potential predators and is sometimes released when the bugs are handled carelessly. The distinctive smell comes from aldehydes such as $CH_3–(CH_2)_2–CH=CH–CHO$, and is chemically similar to pheromones. The nymphs, similar to adults except smaller and without wings, also have stink glands.

The nymphs and adults have piercing mouthparts which most use to suck sap from plants, although some eat other insects. When they group in large numbers they can become significant pests.

The Hitler bug, *Aspongopus janus*

This stink bug is a pest of cucurbits. These red and pale-brown bugs suck the sap from the leaves and tender shoots and devitalize the plants.

Milkweed bugs

The *Lygaeidae* are a family in the Hemiptera (true bugs), comprising twenty-one genera. The family includes insects commonly known as *chinch bugs* and also some of those known as *seed bugs* (the atypical seed bugs, Rhyparochromidae, were formerly placed herein too). Many of the species feed on seeds, although some are predators of other insects, some feed on sap (mucivory), and others feed on blood (hematophagy).

- Milkweed bugs are colored orange (or orange-red) and black.
- They have a long proboscis which they use to pierce the seed and inject salivary enzymes used to digest their food.
- Legs are delicate.

- Adult milkweed bugs have full grown wings which cover the abdomen.

Milkweed bugs often gather in groups on the milkweed plant. This gregarious behavior probably enhances their warning coloration. In the container you may see groups of bugs forming at night. During the day they are usually found feeding on seeds or walking around the container. Sometimes groups form on the side of the container away from the seeds when some of the bugs are molting.

Thrips

Thrips as a group have gained increasing importance over the years with their tendency to inflict significant damage to agricultural and horticultural crops as well as their ever expanding host range resulting in polyphagous habit that exists in the scrub jungle. Thrips are comparatively small-sized insects often referred to as 'fringe wings', 'bladder feet', 'thunderflies' and 'storm flies'. They're mostly phytophagous, mycophagous or predaceous, extending over a wide range of habitats mostly in the tropical, sub-tropical and temperate regions.

Many species of Thrips, *Taeniothrips, Frankliniella, Sericothrips, Caliothrips, Haplothrips* are polyphagous, distributed in various parts of this scrub jungle. The scale of air-borne dispersal being considerable, fluctuations in populations or numerical abundance within a given habitat often results from immigration and migration, as well as through subsequent breeding and mortality. Because of the fringed nature of the wings, thrips are included under the order Thysanoptera (*Thysanos-fringe*) possessing a peculiar asymmetrical 'punch & suck' mouthparts with vestigial right mandible and protrusible bladder at the apex of the tarsi of legs.

Thysanoptera are well-associated with different type of plant formations, and are known to be very susceptible to environmental changes. The main body of the Terebrantia continues to live in the leaf and flower niches, whereas the Tubulifera became adapted to both phytophagous as well as mycophagous habitats, and quite a number of species became associated with formation of plant galls. Some of the Thysanoptera which are classified based on their habitats fall into 3 major categories as given below —

- The Floral Thrips (*Anthophilous thrips*)

- Foliage Surface Dwellers
- The Gall Makers (*Cecidogenous thrips*)

The Thrips species which inhabit the flowers of the plants are commonly called as anthophilous thrips. The seasonal abundance of different species of anthophilous thrips is highly under the influence of the seasonal flowering phenology of the different species of host plants in the campus. Most of the anthophilous species feed and reproduce within the flower. The succession of flowering season also enables several anthophilous species to sustain their population in the scrub jungle of MCC campus. Some of the following species are common inhabiting the flowers of different species of plants in the campus are:

- *Anaphothrips sudanensis* Trybom
- *Frankliniella schultzei* (Trybom)
- *Frankliniella tritici* (Fitch)
- *Megalurothrips distalis* (Karny)
- *Megaphysothrips subramanii* Ramakrishna
- *Mesothrips extensivus* Ananthakrishnan & Jagadish
- *Microcephalothrips abdominalis* (Crawford)
- *Taeniothrips alliorum* Priesner
- *Taeniothrips simplex* (Morrison)
- *Thrips flavus* Schrank
- *Thrips hawaiiensis* (Morgan)

Foliage Surface Dwellers

Monophagy, polyphagy and oligophagy are well reported among the leaf-dwelling thysanopterans. These thrips infest all kinds of leaves but mostly prefer the non-hirsute leaves for feeding. They scrape the surface of the foliage and feed up to the depth of the palisade parenchyma which results in the appearance of silvery-white patches on the adaxial and abaxial sides of the foliage. All life stages can be recorded from the foliage in the case of leaf-dwelling thysanopterans. The population dynamics of these thrips is also highly under the influence of succession of vegetation in the scrub jungle of MCC campus. The following foliage dwellers are reported from the campus:

- *Androthrips flavipes* Schmutz
- *Caliothrips indicus* (Bagnall)

- *Chirothrips maximi* Ananthakrishnan
- *Crotonothrips davidi* Ananthakrishnan
- *Erythrothrips asiaticus* Ramakrishna & Margabandhu
- *Gynaikothrips uzeli* (Zimmerman)
- *Haplothrips ganglbaueri* Schmutz
- *Karnyothrips flavipes* (Jones)
- *Liothrips karnyi* (Bagnall)
- *Liothrips nigripes* (Karny)
- *Lygothrips jambuvasi* (Ramakrishna)
- *Pseudothrips inequalis* (Beach)
- *Sciothrips cardamomi* (Ramakrishna)
- *Scirtothrips citri* Moulton
- *Stenchaetothrips bambusae* (Shumsher)
- *Stenchaetothrips biformis* (Bagnall)
- *Stannardiana variegata* Ananthakrishnan
- *Thilakothrips babuli* (Ramakrishna)
- *Thrips tabaci* Lindemann
- *Tiarothrips subramanii* (Ramakrishna)
- *Toxothrips ricinus* Bhatti
- *Zaniothrips ricini* Bhatti

The Gall Makers (Cecidogenous Thrips)

The gall- forming Thysanopterans are commonly called Cecidogenous Thysanopterans. Gall-formation involves recanalisation and reorientation of plant development resulting in simple or bizarre abnormal structures within which generally, thrips and other insects live, feed and breed. Galling generally involves the ingestion of the plant material by the insects, its concentration or modification in the insect, and the injection of the newly processed material back into the host tissue.

An overview of the gall-forming traits among the Thysanoptera indicates their capacity to exploit the host plants more efficiently and besides being host specific in terms of food requirements, the gall systems actively function as specific breeding sites culminating in specificity of the niche as well.

The identical nature of 'gall-form' irrespective of the geographical variations underlines the consistency in the behavioural patterna in gall thrips in relation to host selection and host affinity. Some of the gall-forming thysanoptera are listed below —

- *Arrhenothrips ramakrishnae* Hood
- *Dolicholepta inquilinus* Ananthakrishnan

Cockroaches

Cockroaches (or simply "roaches") are insects of the order *Blattaria.* This name derives from the Latin word for "cockroach", *Barata.* There are about 4,000 species of cockroach, of which 30 species are associated with human habitations and about four species are well known as pests.

Cockroaches live in a wide range of environments around the world. Pest species of cockroaches adapt readily to a variety of environments, but prefer warm conditions found within buildings. Many tropical species prefer even warmer environments and do not fare well in the average household. The spines on the legs were earlier considered to be sensory, but observations of their locomotion on sand and wire meshes has demonstrated that they help in locomotion on difficult terrain. The structures have been used as inspiration for robotic legs.

Domino cockroach (*Therea petiveriana*)

This magnificent looking cockroach mimics a warningly coloured tortoise beetle. It is a somewhat slow growing species and the nymphs are substrate dwelling like other Polyphagids. Eggcases are buried and can take up to five months to hatch. Nymphs cannot climb glass but adults, especially males, are capable. The specimen depicted is male, females have less prominent antennae.

Praying Mantis

It is so known because of the typical "prayer-like" stance. Praying mantises are exclusively predatory and their diet usually consists of living insects; larger species have been known to prey on small lizards, frogs, birds, snakes, and even rodents. Most species are known to engage in cannibalism. The majority of mantises are ambush predators, waiting for prey to stray too near. The mantis then lashes out at remarkable speed. Some ground and bark species, however, pursue their prey rather quickly. Prey items are caught and held securely with grasping, spiked forelegs ("raptorial legs"); the first thoracic segment, the prothorax, is commonly elongated and flexibly articulated, allowing for greater range of movement of the front limbs while the remainder of the body remains more or less immobile. The articulation of the head is also remarkably flexible, permitting nearly 300 degrees of movement in some species, allowing for a great range of vision (their compound eyes have a large binocular field of vision) without having to move the remainder of the body. As their hunting relies heavily on vision, they are primarily diurnal, but many species will fly at night.

Stick Insects

As its name suggests, the stick insect resembles the twigs amongst which it lives, providing it with one of the most efficient natural camouflages on Earth. It and the equally inconspicuous leaf insect comprise the Phasmida order, of which there are approximately 3,000 species. Stick insect species, often called walking sticks, range in size from the tiny, half-inch-long (11.6-millimeter-long) *Timema cristinae* of North America, to the formidable 13-inch-long (328-millimeter-long) *Phobaeticus kirbyi* of Borneo. This giant measures over 21 inches (55 centimeters) with its legs outstretched, making it one of the world's longest insects. Females are normally larger than males.

Phasmids generally mimic their surroundings in color, normally green or brown, although some species are brilliantly colored and others conspicuously striped. Many stick insects have wings, some spectacularly beautiful, while others resemble little more than a stump. A number of species have spines and tubercles on their bodies.

Found predominantly in the tropics and subtropics, although several species live in temperate regions, stick insects thrive in forests and grasslands, where they feed on leaves. Mainly nocturnal creatures, they spend much of their day motionless, hidden under plants. Many stick insects feign death to thwart predators, and some will shed the occasional limb to escape an enemy's grasp. Little is known about stick insects, making it difficult to declare the vulnerability of their status in the wild. The pet trade presents a potential threat, along with the popular practice of framing their carcasses, like butterflies.

Grasshoppers

The term grasshopper is taken to represent as many as 9,000 different species of singing, jumping insects in two families of

Orthoptera order. Winged insects, with powerful hind legs and strong mandibles, or mouthparts, they can be categorized into two groups - long-horned grasshoppers (family Tettigoniidae) and short-horned grasshoppers (family Acrididae). It has two pairs of wings - the front pair comprises of rigid wings, while the hind pair has larger, membranous wings (often bright in colour). The insect has three pairs of legs, all of which are used for walking purposes.

Katydids

The family *Tettigoniidae,* known in American English as *katydids* and in British English as *bush-crickets,* contains more than 6,400 species. They are also known as *long-horned grasshoppers,* although they are more closely related to crickets than to grasshoppers. Many tettigoniids exhibit mimicry and camouflage, commonly with shapes and colors similar to leaves.

Tettigoniids may be distinguished from grasshoppers by the length of their filamentous antennae, which may exceed their own body length, while grasshoppers' antennae are always relatively short and thickened. The males of tettigoniids have sound-producing organs (via stridulation) located on the hind angles of their front wings. In some species females are also capable of stridulation. The diet of tettigoniids includes leaves, flowers, bark, and seeds, but many species are exclusively predatory, feeding on other insects, snails or even small vertebrates such as snakes and lizards. Some are also considered pests by commercial crop growers and are sprayed to limit growth. Large tettigoniids can inflict a painful bite or pinch if handled but seldom break the skin.

Earwigs

Earwig is the common name given to the insect order *Dermaptera* characterized by membranous wings folded underneath short leathery forewings (hence the literal name of the order — "skin wings"). The abdomen extends well beyond the wings, and frequently, ends in a pair of forceps-like structures called cerci. The order is relatively small among Insecta, with about 1,800 recorded species in 10 families. Earwigs are quite common globally. There is no evidence that they transmit disease to humans or other animals.

The word *earwig* is derived from the Old English *ēare,* which means "ear", and *wicga,* which means "insect" (*Wicga* is in turn related to *wiggle,* and ultimately to other words implying movement, including *way* and *vehicle,* all from PIE **wegh-.*) The name comes from the old wives' tale that earwigs burrowed into the brains of humans through the ear and therein lay their eggs. Earwigs are omnivores that are predisposed to hiding in warm humid crevices and as such, one may crawl into a human ear canal.

Unlike most insects, earwigs exhibit maternal brood care. Earwig females typically fashion a burrow in a confined space, such as under a rock or under the bark of a tree, where eggs are deposited after which the females remain in the burrow, protecting the eggs from destruction, predation and mould, generally not emerging from their burrows until after the eggs have hatched.

Ring-legged earwigs are a very common insect in the campus. They are usually between 1/2 inch and one inch long. Females are slightly larger than males. These insects are brown to black and have yellow legs with brown rings on them. Their antennae are black with white rings. They have no wings. Earwigs live in gardens and fields. They are nocturnal, so they hide under rocks, logs, and plants during the day.

Paper wasps

Paper wasps are 3/4 inch to 1 inch (2-2.5 cm)-long wasps that gather fibers from dead wood and plant stems, which they mix with saliva, and use to construct water-resistant nests made of gray or brown papery material. Paper wasps are also sometimes called *umbrella wasps,* due to the distinctive design of their nests. Unlike yellow jackets and hornets, which can be very defensive, polistine paper wasps will generally only attack if they themselves or their nest are threatened. Since their territoriality can lead to attacks on people, and because their stings are quite painful and can produce a potentially fatal anaphylactic reaction in some individuals, nests in human-inhabited areas may present an unacceptable hazard.

Most wasps are beneficial in their natural habitat, and are critically important in natural biocontrol. Paper wasps feed on nectar, and other insects, including caterpillars, flies, and beetle larvae, and they are often considered to be beneficial by gardeners.

Bumble bee

A *bumblebee* (or *bumble bee*) is any member of the bee genus *Bombus,* in the family Apidae. There are over 250 known species, existing primarily in the Northern Hemisphere.

Bumblebees are social insects that are characterized by black and yellow body hairs, often in bands. However, some species have orange or red on their bodies, or may be entirely black. Another obvious (but not unique) characteristic is the soft nature of the hair (long, branched setae), called pile, that covers their entire body, making them appear and feel fuzzy. They are best distinguished from similarly large, fuzzy bees by the form of the female hind leg, which is modified to form a corbicula; a shiny concave surface that is bare, but surrounded by a fringe of hairs used to transport pollen (in similar bees, the hind leg is completely hairy, and pollen grains are wedged into the hairs for transport). Like their relatives the honey bees, bumblebees feed on nectar and gather pollen to feed their young.

Weaver ant

Oecophylla smaragdina (common names include weaver ant, green ant, green tree ant, and orange gaster) is a species of arboreal ant found in Asia and Australia. They make nests in trees made of leaves stitched together using the silk produced by their larvae.

Weaver ants may be red or green. In Malaysia they are sometimes mis-called "fire ants" because a colloquial name in Malay is *semut api*; another name is *kerengga* (real fire ants are in a different genus, *Solenopsis*).

Butterflies

Like all Lepidoptera, butterflies are notable for their unusual life cycle with a larval caterpillar stage, an inactive pupal stage, and a spectacular metamorphosis into a familiar and colourful winged adult form. Most species are day-flying so they regularly attract attention. The diverse patterns formed by their brightly coloured wings and their erratic yet graceful flight have made butterfly watching a hobby.

Butterflies comprise the *true butterflies* (superfamily Papilionoidea), the *skippers* (superfamily Hesperioidea) and the *moth-butterflies* (superfamily Hedyloidea). Butterflies exhibit polymorphism, mimicry and aposematism. Some migrate over long distances. Some butterflies have evolved symbiotic and parasitic relationships with social insects such as ants. Butterflies are important economically as agents of pollination. In addition, a few species are pests, because they can damage domestic crops and trees in their larval stage.

Of the 19 species of Family Papilionidae recorded in South India, 9 occur in our Campus. There is at least one record of finding the largest Indian butterfly, the common Bird-wing on a stormy day in September 1963. Being the largest and most brilliantly coloured among the Indian butterflies, they are easy to recognise. The third largest Indian butterfly, The Blue Mormon (*Papilio polymnestor*) can be seen in flight at this time of the year. With its subtle blue colour and large size and bird-like flight, The Blue Mormon must be considered as one of our most beautiful campus butterflies.

The Lime Butterfly (*Papilio demoleus*) is very beautiful with its intricate patterns on the wing. The caterpillars of this species feed on citrus leaves. There are four species of *Graphium* among which the Tailed Jay (*G. agamemnon)* and the Common Bluebottle (*G. sarpedon*) are coloured a beautiful green and blue. These can be seen now, hovering around our *Polyalthia* trees. A few minutes of observation will reveal the places where these butterflies have laid their eggs.

All the above species as well as the Black Mormon (*Papilio polytes*) feed on plants such as *Polyalthia, Citrus, Atalantia, Glycosmis, Toddalia and Murraya.* Presumably, the chemicals found in these plants do not protect the caterpillars or the larvae from predators. But two of our common Swallowtails, the Crimson Rose (*Polydorus hector*) and the Common Rose (*P. aristolochiae*) feed on the poisonous plant *Aristolochia indica* during their larval stages. Outside the campus, the caterpillars also feed on a related species, A. *bracteata.* This is supposed to make the adults distasteful to predators. The females of the Black Mormon, whose caterpillars feed only on members of the lime family, mimic the Crimson or Common Rose (Fig. 3 & 4). This phenomenon of Batesian mimicry in these species offers a great opportunity for further explorations. The Black Mormons can be distinguished from the Crimson or Common Rose by the body colour. The former possesses black bodies while the latter are red-bodied.

Papilio demoleus

The *Common Lime Butterfly* (*Papilio demoleus*) is a common and widespread Swallowtail butterfly. It gets its name from its host plants which are usually citrus species such as the lime. Unlike most swallowtail butterflies it does not have a prominent tail. Other names for the butterfly include the *Lemon Butterfly*, the *Small Citrus Butterfly*, the *Chequered swallow tail* and the *Dingy swallow tail.*

This butterfly is very common. It is perhaps the most widely distributed swallowtail in the world. This butterfly is an avid mud-puddler and visitor of flowers. It basks with its wings held wide open on tufts of grass, herbs and generally keeps within a metre above the ground, even on cloudy days. It relies on its quick flight for escape. It is an interesting butterfly in that it has a number of modes of flight. In the cool of the morning, the flight is slow considering that it is an edible and unprotected swallowtail. As the day progresses, it flies fast, straight and low. In the hotter part of the day, it may be found settling on damp patches where it will remain motionless, except for an occasional flutter of wings, if not disturbed. It is also a frequent visitor of flowers in gardens, where it shows a preference for flowers of smaller herbs rather than larger plants such as the ubiquitous *Lantana* with its plentiful blooms. It can be found swarming in the groves of its foodplants.

Euploea core core

The *Common Crow* (*Euploea core*) is a butterfly found in South Asia. In India it is also sometimes referred to as the *Common Indian Crow*, and in Australia as the *Australian Crow*. It belongs to the Crows and Tigers subfamily of the Nymphalidae (Brushfooted butterflies).

The Common Crow is the most common representative of its genus *Euploea*. Like the Tigers (*Danaus* spp.), the Crows are inedible and thus mimicked by other Indian butterflies (see Batesian mimicry). In addition, the Indian species of the *Euploea* genus shows another kind of mimicry, Müllerian mimicry. Accordingly, this species has been studied in greater detail than other members of its genus in India.

The Common Crow (*Euploea core core*) is a glossy black butterfly with brown underside with white marks along the outer margins of the wing. The wingspan is about 8-9 cm and the body also has prominent white spots. On hot days large numbers of these butterflies can be seen mud-puddling on wet sand. This butterfly also gathers on damaged parts of plants such as *Crotalaria, Heliotropium* to forage for chemicals precursors to produce pheromones. Along with other Danaids such as the Tigers, the Common Crow is one of the most common migrating butterfly species. Males and females in equal proportions have been seen to migrate. On hot days large numbers of these butterflies can be seen mud-puddling on wet sand. This butterfly also gathers on damaged parts of plants such as *Crotalaria, Heliotropium* to forage for chemicals precursors to produce pheromones. Along with other Danaids such as the Tigers, the Common Crow is one of the most common migrating butterfly species. Males and females in equal proportions have been seen to migrate.

Danaus plexippus

The *Monarch* (*Danaus plexippus*) is a milkweed butterfly (sub family Danainae), in the family Nymphalidae. The Monarch's wingspan ranges from 8.9–10.2 cm (3½-4 in.). The upper side of the wings is tawny-orange, the veins and margins are black, and in the margins are two series of small white spots. The fore wings also have a few orange spots near the tip. The underside is similar but the tip of the forewing and hindwing are yellow-brown instead of tawny-orange and the white spots are larger.

Monarchs are especially noted for their lengthy annual migration. The length of these journeys exceeds the normal lifespan of most monarchs, which is less than two months for butterflies born in early summer. Monarchs are foul-tasting and poisonous due to the presence of cardenolide aglycones in their bodies, which the caterpillars ingest as they feed on milkweed. Both forms advertise their unpalatability with bright colors and areas of high contrast on the skin or wings. This phenomenon is known as aposematism. Monarchs share this defense with the similar-appearing viceroy butterfly, in an example of Müllerian mimicry.

Danaus chrysippus

The *Plain Tiger* (*Danaus chrysippus*) or outside Asia, *African Monarch*, is a common butterfly which is widespread in Asia and Africa. It belongs to the Danainae ("Milkweed butterflies") subfamily of the brush-footed butterfly family, Nymphalidae. It is believed to be one of the first butterflies to be used in art. A 3500 year old Egyptian fresco in Luxor features the oldest illustration of this species.

The Plain Tiger can be considered the archetypical danaine of India. Accordingly, this species has been studied with in greater detail than other members of its subfamily occurring in India. This butterfly is perhaps the commonest of Indian butterflies and is a familiar sight to practically everyone on the subcontinent. It flies from dawn to dusk, frequenting gardens, sipping from flowers and, late in the day, fluttering low over bushes to find a resting place for the night.

As usual for diurnal butterflies, this species rests with its wings closed. When basking it sits close to the ground and spreads its wings with its back to the sun so that the wings are fully exposed to the sun's rays.

Moths

Bag-worm moths

They belong to the Family Psychidae and are widely distributed insects. The females are apterous whereas the males are winged. Wings are thinly clothed with imperfect scales and exceptionally large frenulum. Some females are devoid of antennae, mouth-parts and legs. Larvae inhabit cases which exhibit great variety of shape and of materials used in their construction.

Termites

The *termites* are a group of social insects usually classified at the taxonomic rank of order *Isoptera*. Termites mostly feed on dead plant material, generally in the form of wood, leaf litter, soil, or animal dung, and about 10% of the estimated 4,000 species (about 2,600 taxonomically known) are economically significant as pests that can cause serious structural damage to buildings, crops or plantation forests. Termites are major detrivores, particularly in the subtropical and tropical regions, and their recycling of wood and other plant matter is of considerable ecological importance.

Photo A-F. **A.** *Thrips tabaci*, B. *Frankiniella tritici*, C. *Bemisia tabaci*, D. *Nephotettix nigropictus*, E. *Megalurothrip distalis*, F. *Myzus persicae*.

Photo G-P. G. *Carausis morosus;* H. *Poekilocerus pictus;* I. *Therea petiveriana;* J. *Blatta orientalis;* K. *Euploea core core;* L. *Danaus chrysippus;* M. *Papilio demoleus;* N. *Danaus plexippus;* O. *Poltdorus hector;* P. *Lamprophorus tenebrosus.*

As eusocial insects, termites live in colonies that, at maturity, number from several hundred to several million individuals. They are a prime example of decentralised, self-organised systems using swarm intelligence and use this cooperation to exploit food sources and environments that could not be available to any single insect acting alone. A typical colony contains nymphs (semi-mature young), workers, soldiers, and reproductive individuals of both genders, sometimes contai-ning several egg-laying queens. Termites are sometimes called "white ants", though they are unrelated to true ants.

Antlions

Antlions are a family of insects in the order Neuroptera with the scientific name *Myrmeleontidae*; the most well-known genus is *Myrmeleo*. There are about 2,000 species.

Depending on species and where it lives, the larvae will either hide under leaves or pieces of wood, in cracks of rocks, or dig pits in sandy areas. Antlion larvae are unusual among the insects as they lack an anus. All the metabolic waste that is generated during the larval stage is stored and is eventually emitted as meconium by the newly eclosed adult.

Fireflies

Lampyrid beetles are commonly called *fireflies* or *lightning bugs* for their conspicuous crepuscular use of bioluminescence to attract mates or prey. Fireflies are capable of producing a "cold light", containing no ultraviolet or infrared rays. This chemically-produced light, emitted from the lower abdomen, may be yellow, green, or pale red in color, and has a wavelength from 510 to 670 nanometers.

Many species can be found in marshes or in wet, wooded areas where their larvae have abundant sources of food. These larvae can also emit light and are often called "glowworms", particularly in Eurasia. The Indian Glow worm, *Lamprophorus tenebrosus* has been studied from the MCC campus.

Dung beetles

Dung beetles are beetles that feed partly or exclusively on feces. All of these species belong to the superfamily Scarabaeoidea; most of them to the subfamilies Scarabaeinae and Aphodiinae of the family Scarabaeidae. This beetle can also be referred to as the scarab beetle. As most species of Scarabaeinae feed exclusively on faeces, that subfamily is often dubbed *true dung beetles*. There are dung-feeding beetles which belong to other families, such as the Geotrupidae (the *earth-boring dung beetle*). The Scarabaeinae alone comprises more than 5,000 species.

Many dung beetles, known as *rollers*, are noted for rolling dung into spherical balls, which are used as a food source or brooding chambers. Other dung beetles, known as *tunnelers*, bury the dung wherever they find it. A third group, the *dwellers*, neither roll nor burrow: they simply live in manure. They are usually attracted by the dung burrowing owls collect.

Bark beetles

Bark beetles are so-named because the best known species reproduce in the inner bark (living and dead phloem tissues) of trees. Some species attack and kill live trees. Most, however, live in dead, weakened, or dying hosts. Bark beetles are ecologically and economically significant. Even outbreak species can help to renew the forest by killing old trees. Other species aid in the decomposition of dead wood. However, several outbreak-prone species are known as notorious pests.

Bark beetles often attack trees that are already weakened by disease, drought, smog, conspecific beetles or physical damage. Healthy trees may put up defenses by producing resin or latex, which may contain a number of insecticidal and fungicidal compounds that can kill or injure attacking insects, or simply immobilize and suffocate them with the sticky fluid. Under outbreak conditions, the sheer number of beetles can however overwhelm the tree's defences, and the results can be disastrous for the lumber industry.

Discussion

Ecosystems always change and are never in total equilibrium. Inevitably they change in composition and in species composition. Since they have a large number of components and interactions, many become adapted to altered circumstances. As such, there is considerable diversity in eco-system behavior and accompanying species diversity. Environmental patchiness favours the persistence of a particular species. Physiologically sensitive species and immobile species at the eco-system edges would be under stress. Insects are rapidly losing their habitats, because they are exposed to climate changes. Global warming puts a new perspective in tropical eco-system where the genetic and physiological plasticity do not allow for adjustment or adapting to changed conditions. For many insects, the pattern of vegetation is an important determinant for the survival of the insect fauna. Some insects creeps like the Lycaenid butterflies which are sensitive to the 'be slow' conditions and mutual interactions become restricted to limited areas. Species of *Theria petiveriana, Anthia sexguttata,* praying mantids and phasmids have less possibility of adapting to environmental perturbations. It has to be emphasized that insects vary enormously in their response to disturbance, some insects locating and benefiting from disturbed patches while others are destroyed by the disturbances.

Decline in population levels of the smaller species like whiteflies and thrips are not uncommon. In general, it may be stated that insect species differ considerably in the constancy of their population levels. A species may be rare in one area while the others like grass hoppers tend to migrate. Identification of the diversity of biotopes and landscapes in essentially important for the diversity of insects and a need for continued investigations on changing faunal trends in scrub jungle vegetation cannot be over emphasized.

References

Annie Phillips 2005 Systematics and diversity of Whiteflies (Hemiptera: Insecta) infesting Tambaram flora. A thesis for Ph.D. Degree awarded by the University of Madras, pp. 288.

Archana Devi 2006 Studies on the Biology of the Ring Legged Earwigs, *Euborellia annulipes* (Anisolabidoidea: Dermaptera). A thesis for M.Phil. Degree awarded by the University of Madras, pp. 125.

Harwin, L.D. 1988 Edge effects and conservation of biotic diversity. *Conservation Biology* 2 : 330-332.

Jencee Mary 2005 A Study on some Earwigs (Dermaptera: Insecta) of Tamil Nadu. A thesis for M.Sc. Degree awarded by the University of Madras, pp. 54.

Jesudasan, Alexander R.W. and Vasantharaj David, B. 1991 Taxonomic studies on Indian Aleyrodidae (Insecta : Homoptera). *Oriental Insects* **25** : 231-434.

Lawton, J.H. 1983 Plant architecture and diversity of phytophagous insects. *Ann. Rev. Ent.* 28 : 23-29.

Nayar, K.K., Ananthakrishnan, T.N. and David, B.V. 1976, *General and Applied Entomology*. Tata McGraw-Hill Publishing Company Limited, New Delhi, pp. 182–184, 235.

Samways, M.J. 1995 *Insect Conservation Biology*, Chapman & Hall

Strong, D.R., Lawton, J.H. and Southwood, T.R.E. 1984 *Insects on plants : Community patterns and mechanisms*. Blackwell, Oxford

Source: http://www.angelfire.com/oh2/Roaches/

Source: http://animals.nationalgeographic.com/animals/bugs/stick-insect.html

Source:http://lifestyle.iloveindia.com/lounge/facts-about-grasshopper-4985.html

Source: http://en.wikipedia.org.

Chapter - 9

Patterns of Adaptive Radiation and Diversification in Cecidogenous Insects

A. Raman

Charles Sturt University & E H Graham Centre for Agricultural Innovation, PO Box 883, Orange, NSW 2800, Australia

e-mail: araman@csu.edu.au

Introduction

In a discussion on the role of biological diversity in ecosystems, it would be useful first to recall the concept of species, which refers to clusters of seemingly 'similar' individuals that are reproductively isolated and that maintain a well-defined genetic identity in the environments where they feed and reproduce successfully (Mayr, 1963; Claridge *et al.,* 1997a). Selection is an essential mechanism in speciation, but adaptive radiation is the key that unlocks selection by developing 'new' behaviours in organisms. Over time, the factors that regulate the new behaviours become integrated into the genetic constructs of organisms and their specialization eventuates, establishing the species. Populations radiate and specialize to fill ecological niches (Futuyma & Peterson 1985; Thompson 1994; Lushai *et al.,* 2002). Therefore, the metaphor would be to perceive 'adaptive radiation' as the 'cog' that regulates the 'gear system' of specialization, which in turn, either accelerates or decelerates the 'wheel' of evolution. In general, among Animalia, insects are a complex set of organisms, because their behaviours and characteristics are hard to be categorized under one or more overarching patterns. Nonetheless, the feeding and reproductive behaviours of many plant-feeding insect taxa and their groups, from extinct to extant examples, have been shown to synchronize with the life-history strategies of flowering plants, thus demonstrating different dimensions of evolution (Schoonhoven *et al.,* 2005). The vibrancy and vitality of specialization among insects are firmly positioned in an 'effort' to minimize interspecific (Jeffries & Lawton, 1984) and intraspecific (Claridge *et al.*, 1997b, Birkle & Douglas, 1999) competition. The evolutionary events as evident in different groups of insects can be artificially, but convincingly, sequenced on a continuum, with 'ecotypes' (Thomas & Singer, 1998) and 'biotypes' (Eastop, 1973) at one extreme and speciation at the other, through 'races' and 'strains', 'sub-' and 'sibling' species (Avise 2000, Lushai *et al.*, 2002). Plant-feeding insects, for instance, have evolved and diversified, and are continuing to evolve and diversify, with the evolution and diversity that has been occurring, and continues to occur, in plant taxa.

Species diversity in and richness of gall-inducing insects: concepts and patterns

Within the broad context of insect phytophagy, the capability to induce galls is a highly specialized behaviour evident among particular taxa in a few groups. The most striking behaviour among gall-inducing insects is that a majority of them are specialist organisms that display a high level of fidelity to their host plants. Various theories explain the origins of insect phytophagy (Labandeira 1997, 2007); the specialized gall-inducing habit among insects is considered to have arisen from the leaf-mining Diptera (e.g., Agromyzidae) and Microlepidoptera (e.g., Nepticulidae, Stigmellidae) (Raman *et al.*, 2005), which, over time, have acquired 'skills' to seek 'new' food sources in concealed, and therefore, 'protected' environments (Crane & Jarzembowski, 1980). That gall-inducing habit originated from tunnelling Lepidoptera (e.g., Tortricidae) and Coleoptera (e.g., Curculionoidea) is also available as an alternate theory (Scott *et al.*, 1994). What is critical, however, will be to recognize that the gall-inducing insect species are scattered among diverse unrelated orders and they induce galls on plant species of unrelated families, indicating that gall-inducing capability has arisen many times and through different evolutionary routes. Of course, physiological differences in the behaviour of the inducing insect and susceptibility levels of plants also have contributed to the variety of galls we see today all over the earth.

Gall-inducing capability among the extant insects exists in taxa belonging to Thysanoptera, Hemiptera, Coleoptera, Diptera, Lepidoptera and Hymenoptera. As indicated earlier, the ability to induce galls does not occur uniformly in all species of these groups; moreover, gall-bearing plants also do not exhibit any discernible pattern because the galls induced by closely related insect taxa develop on plant taxa of unrelated families. In the temperate latitudes, most of the insect-induced galls occur on species of Fagaceae, Salicaceae, Convolvulaceae, Malvaceae, Capparidaceae, Rosaceae, Asteraceae, and Euphorbiaceae (Meyer 1987). In the tropical (and subtropical) world, species of Anacardiaceae, Asteraceae, Combretaceae, Dipterocarpaceae, Fabaceae, Lauraceae, Moraceae, and Myrtaceae host the greatest numbers (Docters van Leeuwen–Reijnvaan & Docters van Leeuwen 1926, Mani 1964, Yukawa & Masuda 1996, Raman *et al.,* 2005b). Richness of gall-inducing insect species, in a global context, is estimated at 133,000 (Espírito–Santo & Fernandes 2007). These insects are highly evolved groups with sophisticated biology and physiology that enable them to utilize their host plants more efficiently and resourcefully than their non-gall-inducing relatives (Raman *et al.,* 2005a, Schaefer *et al.,* 2005, Shorthouse *et al.,* 2005). Although a majority of the gall-inducing insects remain committed to particular plant species (almost conforming to the dictum: one insect species—one plant species), and even to specific plant organs (Abrahamson *et al.,* 1998, Raman *et al.,* 2005a), in recent years, exceptions are being found that induce galls on plant species closely related to their established hosts (Raman 1996, Raman *et al.,* 1996, Gagné 2004, Wool 2005, Raman 2007a). In Cecidomyiidae, for example, although most of gall-inducing gall midges are either monophagous or narrowly oligophagous (Gagné 2004), some taxa of Asphondyliina and Schizomyiina have been found to display polyphagy, indicating that they exhibit trends of shifting to either new or related hosts (Yukawa *et al.,* 2003, Uechi *et al.,* 2002, 2004, Tokuda *et al.,* 2005).

Radiation among gall-inducing Thysanoptera, based on investigations made in the Indian subcontinent (Ananthakrishnan & Raman 1989, Ananthakrishnan 1992) and in Australia (Mound 1994, Mound & Kranz 1997), indicates the Phlaeothripinae to include the maximum numbers of gall-inducing taxa. Out of the nearly 2400 species known in the world taxa of Phlaeothripinae, nearly 40% of them induce galls of varied shapes and sizes, from simple crinkles on leaves to complex pouch galls involving both shoot terminals and juvenile leaves. Because gall-inducing capability occurs in a majority of taxa of *Liothrips* lineage and in a minority of *Phlaeothrips* lineage, gall-inducing capability in the Phlaeothripinae is not monophyletic (Mound & Morris 2005). Because almost half of the number of genera of the *Liothrips* lineage are monobasic and include about 300 species throughout the world, Mound & Morris (2005) consider them not of a single clade. Nonetheless, available information indicates that gall-induction capability has arisen many times, independently, in *Gynaikothrips*, *Kladothrips*, *Liothrips*, and *Teuchothrips*. Within the phlaeothripine taxa, the range of gall types induced is impressively varying: from simple leaf fold and roll galls (e.g., galls on the leaves of *Ficus microcarpa* [Moraceae] induced by *Gynaikothrips ficorum*, with a distribution almost on landmasses of the Gondwana origin) to rosette galls (e.g., on the shoot meristems of *Acacia leucophloea* induced by *Thilakothrips babuli*, restricted to eastern peninsular India) and 'up-to-6-cm' wide pouch galls [housing nearly 100,000 adult and larval thrips in a gall] and on the shoot terminals of *Calycopteris floribundus* [Combretaceae] induced by *Austrothrips cochinchinensis*, confined to western peninsular India). The gall-inducing thrips taxa on different species of *Acacia* explored for possible patterns of adaptive radiation reveal the following (Morris *et al.,* 2002): although nearly 250 species of thrips are known to induce galls on nearly 100 species of *Acacia*, gall-inducing thrips are restricted to two of the three lineages of Australian phyllodinous *Acacia*; the *Acacia*-associated thrips species have radiated in consequence to two unrelated factors: thrips seeking refuge from the harsh arid environment and on-going speciation in *Acacia* (Crespi *et al.,* 2004).

Evolution of sympatric host races in the naturally occurring gall-inducing tephritid populations of North America and Europe explains patterns of host shifts; such shifts occur either by changing the preferences for specific feeding and/or oviposition sites or by developing 'new' physiological adaptations to new host plants or through assortative mating (Abrahamson & Weis 1997, Korneyev *et al.,* 2005). Although the Tephritidae are classified in Diptera, the galls induced by them are not comparable – in terms of variety and complexity – to those induced by their allies, the Cecidomyiidae. The Tephritidae, in general, induce galls on axial organs of plants (e.g., stems and roots), analogous to the 'swollen' tunnels (≈ galls) induced by Coleoptera and Lepidoptera (Abrahamson & Weis 1997, Korneyev *et al.,* 2005). The Cecidomyiidae, on the contrary, induce a variety of galls, mostly on leaves, fewer on flowers, and still fewer on stems (Gagne 2004, Yukawa & Rohfritsch 2005). Among the more-advanced gall-inducing Cynipoidea, differences in the temporally regulated flowering and leafing phenologies in susceptible host plants have regulated the cynipid (Hymenoptera) populations that induce complex and remarkably symmetrical galls by isolating them; divergence and diversification by genetic drift are the mechanisms that have driven the evolution of gall-inducing Cynipoidea (Stone *et al.*, 2001). In spite of such fragmented information on the evolution of different groups of gall-inducing taxa, the present general consensus is that host shifts and radiation in gall-inducing insects are more complex than what seems to be occurring among the non-gall-inducing, plant-feeding insects. Such a complexity is due to composite relationships of gall-inducing insects with their hosts and the spread of gall-inducing insects through different biogeographical realms, which is, in principle, directed by the abundance and variety of plant species (Raman *et al.*, 2009). For instance, patterns of radiation among the gall-inducing Cecidomyiidae (Diptera) are more complicated than what is known among the gall-inducing Tephritidae (Diptera), because of their complex relationships with their host plants and the range of host plants of the Cecidomyiidae. Gall-inducing Tephritidae usually induce galls on shoot apices, mostly of species of Asteraceae (Korneyev *et al.*, 2005). On the contrary, gall-inducing Cecidomyiidae occur attacking leaves and flowers and on a variety of species belonging to diverse plant families – from herbaceous annuals to arborescent perennials. The way the Cecidomyiidae have diversified in the Japanese archipelago and the Beesoniidae (Hemiptera) in temperate and tropical Asia, explained in the following paragraph, would shed some light on the point raised here.

Most species of *Rabdophaga* have radiated on species of *Salix* (Salicaceae) and some of *Rhopalomyia* on Asteraceae have colonized the Japanese Archipelago after their diversification and establishment in Eurasia; *Daphnephila* has come into Japan from tropical south-Asia radiating on *Machilus*; the *Fagus*-infesting gall midges have diversified on the relict species of *Fagus* in the eastern Palearctic and some species of the *Hartigiola* have diversified within Japan, and a majority of Japanese *Asphondylia* are closely related (Yukawa *et al.*, 2005). Among the Beesoniidae (Coccoidea) (unique gall-inducing taxa with highly 'unusual' behaviour even among the Coccoidea – see Gullan *et al.*, 2005), *Beesonia dipterocarpi* and *B. shoreae* occur on *Dipterocarpus tuberculatus* and *Shorea curtisii* (both Dipterocarpaceae) in tropical Asia (Myanmar [Burma] and Singapore), respectively, whereas *B. napiformis* and *B. brevipes* are always associated with different species of *Quercus, Castanopsis,* and *Lithocarpus* of Fagaceae in temperate Asia (parts of China, Japan, and higher elevations of Nepal). Other tropical Beesoniidae, viz., *Gallacoccus secundus, G.. anthonyae, G.. spinigalla, G.. heckrothi, G.. longisetosus, Mangalorea hopeae, Echinogalla pustulata,* and *Danumococcus parashoreae* induce galls on different species of *Hopea, Shorea,* and *Parashorea* (all Dipterocarpaceae) in southern and south-eastern Asia (Gullan *et al.,* 2005; Takagi 1987, 1995, 2001) (Figure 1).

Figure 1: Diversity of galls on species of tropical Dipterocarpaceae induced by Asiatic Beesoniidae (Hemiptera: Coccoidea) A – On *Shorea* sp. induced by *Gallacoccus* sp. B–On *Shorea parvifolia* induced by *Gallacoccus longisetosus.* C–On an unidentified Dipterocarpaceae induced by *Echinogalla* sp. D–On *Shorea falcifera* induced by *Echinogalla pustulata.* E–On *Hopea ponga* induced by *Mangalorea hopeae.* (Courtesy: Professor Sadao Takagi, Sapporo, Japan).

These examples amply demonstrate that the Beesoniidae have primarily evolved with Dipterocarpaceae in the tropics and their association with temperate Fagaceae is a secondarily evolved trait, consequent to their diversification from tropical Asia into temperate eastern Asia (Takagi 2007). Beesoniidae diversification seems to have been strongly influenced by the richness of species in Dipterocarpaceae (*c.* 650 species in 17 genera) in tropical Asia and Fagaceae (*c.* 900 species in eight genera throughout the world, and nearly 700 species in temperate Asia) in temperate Asia, similar to the pattern of diversification of Eriococcidae (Hemiptera: Coccoidea) on highly speciose *Eucalyptus* in Australia (Cook & Gullan 2004). The diversification patterns in the named examples are supported by the apparency theory (Feeny 1976, Strong *et al.,* 1984). Nevertheless, it can be said, in short, that adaptive radiation through host shifts and subsequent speciation, among gall-inducing insects is directed, to a large extent, by host-plant traits (e.g., host-plant chemistry) and competitive interactions of the gall inducer in selecting the 'best' host resources, and to a lesser extent through windows of escape from natural enemies (Abrahamson *et al.,* 1994). In spite of this level of understanding, a well-articulated list of reasons why specific host shift happens among gall-inducing insects, especially because host shifts seem to be occurring only among related plants, is hard to construct.

Figure 2: (A–D) Galls involving oppositely placed leaflets on four southern-Indian species of *Acacia* (Mimosaceae) induced by four species of *Contarinia* (suspected *Acacidiplosis*) (Diptera: Cecidomyiidae). D – Cylinder–piston galls induced on *Acacia ferruginea* by *Contarinia* sp. (=*Lobopteromyia ramachandrani*; suspected *Acacidiplosis*) (Courtesy: Dr. S. Amerjyothy, Madras, India)

Richness, host relations, and distribution of gall-inducing insect species in the Indian subcontinent

Gall-inducing ability among the known insects of the Indian subcontinent occurs mostly in species of Thysanoptera, Hemiptera, Diptera, and Hymenoptera, and to a lesser extent in species in Lepidoptera and Coleoptera (Mani 1973, 2000, Raman 2007a, 2007b), a pattern similar to what exists in the remainder of the world (Dreger-Jauffert & Shorthouse 1992, Espírito–Santo & Fernandes 2007, Mendonça 2007). Within these orders, gall-inducing taxa are confined to particular families or subfamilies. For example, among the gall-inducing Psylloidea of the Indian subcontinent, only Triozidae includes nearly 25 established species on plants belonging to Combretaceae, Ebenaceae, Ericaceae, Euphorbiaceae, Lauraceae, Moraceae, Myrtaceae, Sabiaceae, Salicaceae, and Urticaceae (Yang & Raman 2007). Close to 2000 gall types and almost an equal number of gall-inducing insects have been reported in the Indian subcontinent (Mani 2000, Raman 2007a). In general, dicotyledons host a majority of gall-inducing insects. Among the remainder of vascular plants, only *Pinus* and *Picea* (Pinaceae) from the Himalaya, *Gnetum ula* (Gnetaceae) in peninsular India (Mani 2000), and a few species of grasses (Mani 1973, Gagné 1985, Grover 1988) bear galls induced by Diptera (Cecidomyiidae). In terms of numbers, galls occur unevenly in Fabaceae, followed by Anacardiaceae, Asteraceae, Combretaceae, Cucurbitaceae, Lauraceae, and Moraceae, which host the highest number of gall-inducing insect species, whereas Capparidaceae, Convolvulaceae, Euphorbiaceae, Fagaceae, Malvaceae, Rosaceae, and Salicaceae, which bear galls in temperate parts of the world, do not host gall-inducing insect species in comparable numbers (Mani 2000). A caveat, however, remains that several insects and gall-bearing plants are yet to be documented and enumerated (Raman 2007b).

Distribution of gall-inducing insects in the Indian subcontinent shows various peculiarities compared with the patterns of distribution of these insects in the remainder of the world. Gall-inducing aphids, cynipids, and sawflies generally occur only along the cold Himalayan slopes, whereas gall-inducing thrips and coccids occur only in warm, humid peninsular India (Raman 2007a). Although no gall-inducing psylloids have been recorded from the Indian Himalaya (Mani 2000), gall-inducing *Cecidopsylla schimae* (Hemiptera: Calophyidae) and a few species of Triozidae associated with *Symplocos* occur in the Nepalese Himalaya (Raman *et al.,* 2009). *Trioza jambolanae* (Hemiptera: Triozidae), which induces protruding globular galls on the leaves of *Syzygium cumini* (Myrtaceae) (Mathur 1975, Kandasamy 1986, Raman *et al.,* 1996) occurs throughout the Indian subcontinent, irrespective of variations in latitudinal and longitudinal gradients, although its life-history traits (e.g., number of generations per year) vary between the populations that occur in peninsular India (Raman 1991) and in the Himalayan foothills (Mathur 1935, 1946). In contrast, the sea-urchin shaped galls on *Hopea ponga* (Dipterocarpaceae) induced by *Mangalorea hopeae* (Hemiptera: Beesoniidae) occur only along the western coast of the peninsula (Raman & Takagi 1992). Several other species of gall inducers occur only in specific biogeographical subregions: a majority of gall-inducing insects in peninsular India is endemic to this region; however, some of them in the central-Indian plateau have been documented to display morphological affinities to elements of the Afrotropical and Mediterranean biogeographical regions (Mani 2000, Raman 2007b). On the contrary, those in the western and north-western Himalaya include a high proportion of European and central–Asian elements (Mani 2000). Although the channeling effect of the Himalaya on monsoon currents and the topographic peculiarity of the peninsula explain the oddities in the distribution of gall-bearing plants and gall-inducing insects in India (Raman 2007b), why gall-inducing sawflies, cynipids, and aphids occur only along the Himalayan slopes, and why most of the gall-inducing thrips and coccids occur only in the tropical stretches of peninsular India, are awaiting answers (Raman 2007a).

Indian gall flora includes galls of inimitable morphologies; the most unique among them is the cylinder–piston gall, that involves two oppositely lying leaflets of *Acacia ferruginea* (Mimosaceae), induced by a species of *Contarinia* (= *Lobopteromyia ramachandrani*; presently suspected to be a species of *Acacidiplosis*, K M Harris, personal communication) (Rohfritsch 1971, Mani 2000). Dr. Saminathan Amerjyothy (Presidency College, Madras, India) has found galls of morphology similar to those of *A. ferruginea* on three other taxa of *Acacia* in the eastern peninsular India (scrub vegetation), all of which involve two oppositely lying leaflets in organizing a gall; therefore these are also being suspected to be induced by species of *Contarinia* (or *Acacidiplosis*) (Figure 2).

Gall-inducing insects demonstrate a high degree of orientation to specific trophic niches (Raman 1996). Close to 90% of them are host specific (Meyer 1987), displaying a high level of fidelity to particular species of plants. Importantly, gall-inducing insects are specific to particular plant organs, i.e., a leaf-gall inducing insect will induce galls only on leaves. The relationship between different species of *Crotonothrips* (Thysanoptera: Phlaeothripidae) and those of *Memecylon* (Melastomaceae) distributed almost evenly in the dry-scrub (southern stretches of the Eastern Ghats) and dry-deciduous (eastern slopes of southern Western Ghats) vegetation of peninsular India would illustrate the point argued here: *C. gallarum*, *C. coorgensis*, *C. dissimilis*, *C. memecylonicus*, and *C. dantahasta* induce galls on *Memecylon* sp., *M. talbotianum*, *M. lawsonii*, *M. lushingtonii*, and *M. edule*, respectively, and galls arising in these interactions range from simple epiphyllous rolls to complex leaf rosettes (Ananthakrishnan 1976, Raman & Ananthakrishnan 1985). No *Crotonothrips* occur in either the remainder of India or the rest of the Orient. On the contrary, gall induction on leaflets of different species of *Schefflera* (Araliaceae) by diverse species of *Liothrips* (Thysanoptera: Phlaeothripidae) is known, not only from cooler and upper slopes (above 1500 m asl) of the Western Ghats, peninsular India (Ananthakrishnan 1972, Krishnamurthy *et al.,* 1975), but also in rain forests of Northeast India (e.g., Thoubal and Chandel districts, Manipur), Java and Celebes (Indonesia) (Karny *et al.,* 1913), and the Philippines (Uichanco 1919). However, exceptions such as *Thilakothrips babuli* (Thysanoptera: Phlaeothripidae) and *Procontarinia mangiferae* (Diptera: Cecidomyiidae) occur. *T. babuli* usually induces rosette galls on vegetative shoot meristems of *Acacia leucophloea* (Mimosaceae) (Raman & Ananthakrishnan 1983); however, rarely, *T. babuli* induces galls on floral meristems. *Procontarinia mangiferae* induces galls not only on the leaves, but also on the stems of *Mangifera indica* (Mani 2000, Raman *et al.,* 2009). The capability of *T. babuli* to shift gall-sites on the same host plant is because of its adaptive capability to avoid the nonavailability of vegetative meristems during non-monsoon seasons (Varadarasan & Ananthakrishnan 1982). The level of sensitivity of gall-inducing insects is so high that they can discriminate between 'subtle' differences in host-plant physiology, chemistry, development and phenology, and go for their most preferred organs on specific host plants. Gall-inducing insects can 'skillfully' distinguish even between plant hybrids

(Floate & Whitham 1995), mainly because of the secondary plant metabolism and the volatiles produced. However, we need to be aware that a few taxa such as *Dasineura marginemtorquens* (Diptera: Cecidomyidiae) and *Adelges abietis* (Hemiptera: Adelgidae) exist; *D. marginemtorquens* is capable of attacking non-host plants, *A. abietis* is capable of ovipositing on resistant plants.

The imperative host dependence underscores the 'ability' of a majority of gall-inducing insects to choose the right host plant and right site at the right time for oviposition initially and feeding subsequently: gravid females of *Trioza jambolanae* (Hemiptera: Triozidae) oviposit along the lateral sutures of leaves of *Syzygium cumini* (Myrtaceae), displaying an extraordinary level of precision; oviposition by *T. jambolanae* coincides with emergence of new-leaf flushes after monsoon rains in natural areas near Madras (Raman 1991). Even in less-complex gall-inducing taxa such as *Acythopeus burkhartorum* (Coleoptera: Curculionidae), the gravid female displays an extraordinary level of sophistication in its oviposition behaviour (Raman *et al.,* 2007). Because the inducing insect (and in some instances, also its progeny) spends a major part of its life cycle within the gall, the gall-inducing insects appear to indicate that they take no 'chances', given that they invariably encounter diverse species of plants in a heterogeneous natural environment. To achieve this, these specialist insects use tactics that are more sophisticated than their non-gall-inducing relatives by directing their movements towards their preferred host plants and sites (Raman 1987).

Evolutionary pressure between random search and specific pattern of orientation in host-plant selection by phytophagous insects depends on a delicate balance between cost in time and energy for finding the host and difficulty in detecting or processing information from the host plant. Consequently, gall-inducing insects that remain restricted to a single plant species have a need to search and locate the most 'ideal' host plant within specific temporal constraints. The plant that offers least resistance becomes the host; under extremely rare circumstances, gall-inducing insects also invade plants closely related to their most-preferred hosts and survive on them by inducing galls. *Trioza fletcheri minor* (Hemiptera: Triozidae) induces galls on leaves of *Terminalia tomentosa* and *T. arjuna* (Combretaceae), common not only in the temperate foothills of the

Himalaya and the Indo-Gangetic plains (Crawford 1912, Mathur 1935), but also in the subtropical and deciduous forests in peninsular India (Mani 1973). Through karyotyping, populations of *T. f. minor* obtained from both *T. tomentosa* and *T. arjuna* have been established as the same species (Raman *et al.*, 1997). *Trioza f. minor* is also known to occur and induce galls on *T. catappa*, *T. paniculata*, and *T. tomentosa* (Mathur 1975). Whereas almost all species of Triozidae – a group that principally includes gall inducers – are restricted to specific host-plant species, the behaviour of *T. f. minor* is exceptional, because it induces galls not only on the leaves of several species of *Terminalia*, but also on the flowers of more than one species of *Terminalia* (Sokhi & Kapil 1984, 1985). Patterns of occurrence of *T. f. minor* in naturally occurring populations of different species of *Terminalia* could be due to poor configuration of the taxonomy of *Terminalia* – which in reality could be either sympatric or geographic races, rather than well defined species. Alternatively, incidence of *T. f. minor* in the naturally occurring populations of *Terminalia* could also be due to evolution of host races in *T. f. minor*, which would isolate themselves as individual species in time. In short, what is evident, as of now, is that a high degree of specialization and diversification is ongoing between gall-inducing insects and their host plants.

Conservative radiation in the Cecidomyiidae living on Indian Anacardiaceae

As a case study, I here discuss patterns of radiation that are apparent in the Cecidomyiidae that induce galls on different taxa of Indian Anacardiaceae. Globally, Anacardiaceae includes 80 genera and about 750 species (Pell 2004), whereas in India, this natural order includes 23 genera and 68 species (Chandra & Mukherjee 2000). Most species of Anacardiaceae are confined to the tropics and their species diversity is particularly high in Malesia (Mukherjee 1997), the phytogeographic region between the southern Malay Peninsula and eastern New Guinea. Anacardiaceae is estimated to have originated in the Palaeocene (65–55 mya) (Muller 1984) in the Gondwana (Gentry 1982). Among the extant Anacardiaceae, a variety of perennial plant forms exist, and all species include a characteristic endowment of latex and resins, besides a range of simple and complex phenols and esters (Aguilar-Ortigoza *et al.*, 2003). The catechols, resorcinols, and biflavonoids in different members of Anacardiaceae are implicated to play a defence role against vertebrate and insect herbivores (Mitchell 1990). Certain other species of Anacardiaceae contain methyl gallic acid, gallic acid, a methyl ester of 3,4,5–trihydroxybenzoic acid, and 5–(12–heptadecenyl)–resorcinol (an anti-microbial compound) (Cojacaru *et al.*, 1986, Saxena *et al.*, 1994), and ß–ocimene (an anti-insectan compound) (Chen & Wiemer 1984). Mangiferin is a key phenolic glucoside in *Mangifera indica* (Nott & Roberts 1967).

Among the Indian Anacardiaceae, *Holigarna arnottiana* and *Semecarpus anacardium* (Semecarpeae), *Lannea coromandelica* and *Spondias mangiferae* (Spondiadeae) *Buchanania lanzan* and *Mangifera indica* (Anacardieae), *Pistacia integerrima*, *Pistacia khinjuk* and *Rhus semialata* (Rhoeae), and one 'indeterminate' taxon bear galls, mostly on leaves and rarely on stems. Twenty-one gall-midge taxa (Diptera: Cecidomyiidae) induce galls on these plant taxa, with 16 of them on leaves, four on flowers, and one on stems. *Procontarinia mangiferae* is the singular species that induces galls, not only on the leaves, but also on the stems of *M. indica*. *Oligotrophus mangiferae* (an unplaced species in Cecidomyiinae, sensu Gagné 2004, p. 279) is an exceptional, which induces stem galls on *M. indica*. Six established species of psyllids (Hemiptera: Calophyidae), including, the matchless taxon Apsylla cistellata (Apsyllinae), occur on Indian Anacard-iaceae (Mani 2000, Raman, 2007, Raman *et al.*, 2009).

Diversification of Cecidomyiidae on Indian Anacardiacae needs to be viewed against the the Upper Palaeocene record of a cecidomyiid-induced gall by *Procontarinia echinogalliperda* (or its ancestor) on the leaves of *Eomangiferophyllum damlagiri-ensis* (an analogue of the modern *Mangifera indica*) from north-eastern India (Srivastava *et al.*, 2000). Whereas, a majority of *M. indica* infesting modern taxa of Cecidomyiidae induce galls only on leaves, the behaviour of *Procontarinia mangiferae* reported first on the stems of *M. indica* by Mani as *Rabdophaga mangiferae* in 1959, and later on the leaves of *M. indica* by Grover in 1962 needs a closer re-examination to verify whether *P. mangiferae* has radiated differently to other *M. indica*-infesting Cecidomyiidae and therefore infests both stems and leaves simultaneously, or on the contrary, whether the leaf infestation has preceded stem infestation (or *vice versa*). The capability of multiple shifts has been implicated in gall-inducing Cecidomyiidae-host plant relationships in a global context (Gagné 2004). However, the limitation of information on the Cecidomyiidae that infest Indian Anacardiaceae restricts making a similar inference. For instance, except for *P. mangiferae*, the 23 recorded taxa of Cecidomyiidae (15 named and eight unnamed taxa) that induce galls of distinctly different morphologies on *M. indica* (Raman *et al.*, 2009) remain committed to their respective specific organs, viz., leaves and flowers. The Indian Anacardiaceae-infesting Cecidomyiidae, therefore, can be inferred to be remaining conservative, without displaying any worthwhile dimension of adaptive radiation. *Holigarna,* a southern-Indian genus of Anacardiaceae, has diversified with five well-defined species, viz., *H. arnottiana*, *H. beddomei*, *H. ferruginea*, *H. grahami*, and *H. nigra* in the wet, tropical south-western peninsula. An unnamed Cecidomyiidae induces leaf galls on *H. arnottiana* remains committed to this species and has not diversified onto the other four taxa of *Holigarna*, in spite of geographical proximity of the five species and their endemicity to the subregion. Definite patterns of adaptive radiation in Cecidomyiidae are strikingly evident in species of Asphondyliini and Lasiopterini, which always harbour a fungal associate (a symbiont) in their galls (Yukawa & Rohfritsch 2005, Rohfritsch 2008); such a pattern of adaptive radiation appears to be mediated either by the metabolic products of the fungal symbiont or by its ecological relationship with the insect (Roskam 2005). The taxa that do not harbour the fungal associate in their galls, such as those of the Cecidomyiidae, seem to remain conservative in their host fidelity and strongly committed to their specific hosts and host sites; all the known gall midges that induce galls on *M. indica* are members of the Cecidomyiini (Cecidomyiidae: Cecidomyiidi) (Raman *et al.,* 2009).

Reproductive isolation and temporal isolation steer radiation among plant-feeding insects. Reproductive isolation progresses through ecological specialization of the insect on specific plant species (Caillaud & Via 2000). Temporal isolation, on the other hand, progresses through mechanisms directed by the host plants (Groman & Pellmyr 2000, Joy & Crespi 2007). Reproductive isolation, determined by prezygotic isolation, associated with fidelity of mating on a specific host plant (Berlocher 2000), leads to formation of host-plant races with moderate levels of reproductive isolation leading to diversification into independent species (Abrahamson *et al.,* 2001, Drès & Mallet 2002). Examples illustrating reproductive-isolation mediated diversification involving rapid shifts to related plant taxa occur plentifully among gall-inducing

Eulophidae (Chalcidoidea) (Raman & Withers 2003, La Salle 2005). For instance, in the last decade, gall-inducing Eulophidae have been invading mixed-species plantations of *Eucalyptus* and horticultural stands of *Erythrina* in the Middle-East (Mendel *et al.,* 2004), South Asia (Jacob & Kumar 2009), and the many islands in the Pacific (Yang *et al.,* 2004). Host shifts, although not as striking and 'rapid' as in the Eulophidae, occur in other gall-inducing Hymenoptera, such as the Agaonidae (Cook & Rasplus, 2003) and Cynipidae (Rokas *et al.,* 2003). In spite of being speciose, adaptive radiation in the species of Cecidomyiidae has been demonstrated only in Japanese (Yukawa *et al.,* 2003; Uechi *et al.,* 2002, 2004; Tokuda *et al.,* 2005) and North American Asphondyliini (Joy & Crespi, 2007).

Several species of *Procontarinia* inducing galls on *M. indica* do not show any shift either between organs within *M. indica* or between other anacardiaceous taxa in India, such as *Anacardium, Buchanania, Lannea, Holigarna, Pistacia, Semecarpus*; notably, *Anacardium, Buchanania, Lannea, Holigarna, Pistacia, Semecarpus* serve as hosts to species of Cecidomyiidae other than *Procontarinia.* Because shifts to a different plant species require adaptation to distinctly varied characteristics in the host plant (e.g., morphology, chemistry, and phenology), the shifts within a specific plant taxon from one organ to another need not necessarily involve adaptation of the magnitude what would be needed for a shift from one species to another. Cook *et al.* (2002) have demonstrated that shifts within plant lineages are more common than shifts between plant lineages and have suggested that shifts between organs within a plant are less challenging, given that the host genotype is the same. But the proteomics of organs of the same plant could be different (Hochholdinger *et al.,* 2006), which necessitates the ability of gall-inducing insects to adapt to the local, organ-specific, host-plant proteomics. Therefore, it can be surmised why the *M. indica*-associated *Procontarinia* are conservative in the context of host or organ shifts, lacking either the ability or the need to adapt to organ-specific proteomics of *M. indica.* Moreover, the lack of radiation from leaves to other organs of *M. indica* indicates that neither the feeding behaviour nor the salivary physiology of the respective gall-midge species has changed over time. Conservative radiation characterizes *Procontarinia* biology on *M. indica* in the Indian subcontinent. An inter-simple sequence repeat marker study of 60 Indian and non-Indian *M. indica* varieties shows that the *M. indica* varieties from Australia,

USA, Israel, Taiwan, and Thailand cluster into distinctly separate groups from the group that would broadly include the Indian *M. indica* varieties, thus reinforcing the distinct genetic divergence between the Indian and non-Indian varieties (Pandit *et al.,* 2007). In the context of *Procontarinia*, the genetic divergence of the non-Indian *M. indica* varieties is reflected in *P. matteiana* infestation behaviour in South Africa, because the North American (Florida) varieties 'Keitt', 'Kent' and 'Irwin' grown in South Africa manifest only aborted galls consequent to *P. matteiana* infesta-tion, indicating only partial susceptibility to *P. matteiana*, whereas *P. matteiana* behaviour in India results in the manifestation of well-developed galls of identical morphology on different Indian varieties of *M. indica* (Raman *et al.,* 2009).

Conservative radiation in the Indian Anacardiaceae-infesting Cecidomyiidae is striking when seen against the patterns of relatively liberal radiation of an endemic Indian psyllid *Trioza fltecheri minor* (Hemiptera: Triozidae), which induces galls on the leaves of several species of *Terminalia* (Combretaceae): *T. tomentosa, T. arjuna, T. catappa, T, paniculata,* and *T. tomentosa* x *T. arjuna* (Crawford, 1912, Mathur, 1935, 1975; Mani, 1973; Raman *et al.,* 1996). Populations of these species of *Terminalia* occur fairly uniformly throughout the Indian subcontinent (Champion & Seth, 1968). On the contrary, diverse species of Cecidomyiidae infest different species of *Terminalia* in Peucang, Panaitan, and Krakatau islands of Indonesia (Docters van Leeuwen-Reijnvaan & Docters van Leeuwen, 1926; Yukawa & Partomihardjo, 1997) that are climatologically similar to that of tropical southern India. Relative incidence of Cecidomyiidae on *Terminalia* in Indonesian islands and the relative absence of Cecidomyiidae on *Terminalia* in India indicate that the shifts and consequent radiation of gall-inducing Cecidomyiidae are directed by the host-plant characteristics (Raman, 2007b).

Ecological opportunities for the gall-inducing Indian Cecidomyiidae on different taxa of Anacardiaceae display their capability to induce galls on leaves mostly, and on flowers and stems rarely. Knowledge of the arthropod community associated with different galls induced by the Indian Cecidomyiidae is poor, except that of Yadav & Prasad (1992). How the parasitic Hymenoptera influence the adaptive radiation and evolution of gall midges is well known (Stone & Schönrogge, 2003, Yukawa & Rohfritsch, 2005; Roskam, 2005; Stireman *et al.,* 2006). Obviously, the Indian Cecidomyiidae on Anacardiaceae have been living conservatively in their respective galls, and have evolved specializing minimally along with those of those their host plants.

Conclusion

One key characteristic of gall-inducing insects is their fidelity to particular host plants. In fact, several reviews and experimental investigations point to the fact that the identity of a gall-inducing insect can inform the identity of the host plant, and *vice versa* (Sýlven, 1979; Raman, 1996; Abrahamson *et al.*, 1998, Yang *et al.*, 2002). The possible absence of resistance-breaking genes in Indian Cecidomyiidae explains why these organisms have not radiated and diversified as aggressively as the gall-inducing Eulophidae (Hymenoptera) on different taxa of *Eucalyptus*, introduced into India and many nations in the Middle East as commercial plantations. Alternatively, the host-plant populations are restricting the gene flow between specific gall-midge populations, through their secondary chemistry because, the host-plant mediated impediments on the breeding behaviours impact on the radiation of gall-inducing insects (Martinez *et al.*, 2005). What can be said in conclusion is that a majority of gall-inducing insects of the Indian subcontinent seem to show features of conservative diversification and whatever little has been documented so far, appear to be strongly plant mediated, as discussed in the instance of *Trioza fletcheri minor* (Hemiptera: Triozidae) interacting and inducing galls on different species of *Terminalia*.

Acknowledgement

Jonathan Majer (School of Agriculture & Environment, Curtin University of Technology, Perth, Australia) reviewed the final draft of this manuscript and offered many useful comments and suggestions; I thank Jonathan for his kindness.

References

Abrahamson WG and Weis AE 1997. Evolutionary ecology across three trophic levels: goldenrods, gallmakers, and natural enemies. Princeton, USA: Princeton University Press, Princeton, USA.

Abrahamson WG, Melika G, Scrafford R and Csóka G. 1998. Gall-inducing insects provide insights into plant systematic relationships. *American Journal of Botany* 85, 1159–1165.

Abrahamson WG, Brown JM, Roth SK, Sumerford DV, Horner JD, Hess MD, Tørgerson–How S, Craig TP, Packer RA, and Itami JK 1994. Gallmaker speciation: an assessment of the roles of host-pant characters, phenology, gallmaker competition, and natural enemies. *In* Price PW, Mattson WJ, & Baranchikov YN (Editors) *The ecology and evolution of gall-forming insects.* USDA Forest Service, St Paul, USA. pages 208–222.

Abrahamson WG, Eubanks MD, Blair CP and Whipple AV 2001. Gall flies, inquilines and goldenrods: a model for host-race formation and sympatric speciation. *American Zoologist* 41, 928–938.

Aguilar-Ortigoza CJ, Sosa V, and Aguilar-Ortigoza M 2003. Toxic phenols in various Anacardiaceae species. *Economic Botany* 57, 354–364.

Ananthakrishnan T N 1972. Further studies on Indian gall thrips. *Marcellia* 37, 3–20.

Ananthakrishnan TN 1976. New gall thrips of the genus *Crotonothrips* (Thysanoptera). *Oriental Insects* 10, 411–419.

Ananthakrishnan TN 1992. Unique aspects in the biology of thrips-induced galls. *In* Shorthouse JD & Rohfritsch O (Eds.) *Biology of insect-induced galls.* Oxford University Press, New York, USA. pages 185–194.

Ananthakrishnan TN and Raman A 1989. *Thrips and gall dynamics.* E J Brill, Leiden, The Netherlands.

Avise JC 2000. *Phylogeography: the history and formation of species.* Harvard University Press, Cambridge, Massachusetts, USA.

Berlocher SH 2000. Radiation and divergence in the *Rhagoletis pomonella* species group: inferences from allozymes. *Evolution* 54, 543–557.

Birkle LM and Douglas AE 1999. Low genetic diversity among pea aphid (*Acyrthosiphon pisum*) biotypes of different plant affiliation. *Heredity* 82, 605–612.

Caillaud MC and Via S 2000. Specialized feeding behaviors influences both ecological specialization and assortative mating in sympatric host races of pea aphids. *American Naturalist* 156, 606–621.

Chandra D and Mukherjee SK 2000. Anacardiaceae. *Flora of India* 5, 435—510.

Chen TK and Wiemer DF 1984. A volatile leafcutter ant repellent from *Astronium graveolens. Naturwissenschaften* 71, 97–98.

Claridge MF, Dawah HA and Wilson MR (Eds.) 1997a. *Species: the units of biodiversity.* Systematics Association Special Volume Series 54. Chapman & Hall, London, UK.

Claridge MF, Dawah HA and Wilson MR 1997b. Species in insect herbivores and parasitoids – sibling species, host races and biotypes. *In* Claridge MF, Dawah, HA & Wilson MR (Eds.) *Species: the units of biodiversity.* Systematics Association Special Volume Series 54. Chapman & Hall, London, UK. pages 247–272.

Cojocaru M, Droby S, Glotter E, Goldman A, Gottlieb HE, Jacoby B, and Prusky D 1986. 5–(12–heptadecenyl)–resorcinol, the major component of the antifungal activity in the peel of mango fruit. *Phytochemistry* 25, 1093–1095.

Cook JM, Rokas A, Pagel M, and Stone GN 2002. Evolutionary shifts between host oak species and host plant organs in *Andricus* gall wasps. *Evolution* 56, 1821–1830.

Cook JM and Rasplus JY 2003. Mutualists with attitude: coevolving fig wasps and figs. *Trends in Ecology & Evolution* 18, 241–248.

Cook LG and Gullan PJ 2004. The gall-inducing habit has evolved multiple times among the eriococcid scale insects (Sternorrhyncha : Coccoidea: Eriococcidae). *Biological Journal of the Linnean Society* 83, 441–452.

Crane PR and Jarzembowski EA 1980. Insect leaf miners from the Palaeocene of southern England. *Journal of Natural History* 14, 629–636.

Crawford D L 1912. Indian Psyllidae. *Records of the Indian Museum* 7, 419–437.

Crespi BJ, Morris DC, and Mound LA 2004. Evolution of ecological and behavioural diversity: Australian Acacia thrips as model organisms. Australian Biological Resources Study Canberra and Australian National Insect Collection, Canberra, Australia.

Docters van Leeuwen—Reijnvaan J and Docters van Leeuwen WM 1926. *The zoocecidia of the Netherlands East Indies*. Drukkerij de Unie, Batavia, Indonesia.

Dreger-Jauffret F and Shorthouse JD 1992. Diversity of gall-inducing insects and their galls. *In* Shorthouse JD & Rohfritsch O (Eds.) *Biology of insect-induced galls*. Oxford University Press, New York, USA. pages 8–33.

Drès M and Mallet J 2002. Host races in plant-feeding insects and their importance in sympatric speciation. *Philosophical Transactions of the Royal Society of London (B)* 357, 471–492.

Eastop VF 1973. Biotypes of aphids. *In* Lowe AD (Ed.) *Perspectives in aphid biology*. Bulletin No. 2. The Entomological Society of New Zealand (Inc.), Wellington, New Zealand. pages 40–51.

Espírito–Santo MM and Fernandes GW 2007. How many species of gall-inducing insects are there on earth, and where are they? *Annals of the Entomological Society of America* 100, 95–99.

Feeny P 1976. Plant apparency and chemical defense. *Recent Advances in Phytochemistry* 10, 1–40.

Floate KD and Whitham TG 1995. Insects as traits in plant systematics: Their use in discriminating hybrid cottonwoods. *Canadian Journal of Botany* 73, 1–13.

Futuyma DJ and Peterson SC 1985. Genetic variation in the use of resources by insects. *Annual Review of Entomology* 30, 217–238.

Gagné RJ 1985. A taxonomic revision of Asian rice gall midge, *Orseolia oryzae* (Wood-Mason) and its relatives. *Entomography* 3, 127–168.

Gagné RJ 2004. A catalog of the Cecidomyiidae (Diptera) of the world. *Memoirs of the Entomological Society of Washington* 25, 1–408.

Gentry AH 1982. Neotropical floristic diversity: phytogeographical connections between Central and South America, Pleistocene climatic functions, or an accident of the Andean orogeny? *Annals of the Missouri Botanical Garden* 69, 557–593.

Groman JD and Pellmyr O 2000. Rapid evolution and specialization following host colonization in yucca moth. *Journal of Evolutionary Biology* 13, 223–236.

Grover P 1962. Studies on new gall midges VI (Cecidomyiidae: Nematocera: Diptera) from India. *Proceedings of the National Academy of Sciences of India (B)* 32, 312–318.

Grover P 1988. Cecidomyiid larvae and host-plant interaction. *In* Ananthakrishnan TN & Raman A (Eds.) *Dynamics of insect–plant interaction*, Oxford & IBH Publishing Company, New Delhi, India. pages 177–200.

Gullan PJ, Miller DR, and Cook LG 2005. Gall-inducing scale insects *In* Raman A, Schaefer CW, & Withers TM (Eds.) *Biology, ecology, and evolution of gall-inducing arthropods.* Science Publishers, New Hampshire, USA. pages 143–158.

Hochholdinger F, Sauer M, Dembinsky D, Hoecker N, Muthreich N, Saleem M, and Liu Y 2006. Proteomic dissection of plant development. *Proteomics* 6, 4076–4083.

Jacob JP and Kumar AR 2009. Incidence of Galls Induced by *Leptocybe invasa* (Hymenoptera: Eulophidae) on Seedlings of *Eucalyptus camaldulensis* and *E. tereticornis* (Myrtaceae) from Different Seed Sources in Southern India *International Journal of Ecology and Environmental Sciences* 00, 000–000 (in press).

Joy JB and Crespi BJ 2007. Adaptive radiation of gall-inducing insects within a single host-plant species. *Evolution* 61, 784–795.

Kandasamy C 1986. Taxonomy of south Indian psyllids. *Records of the Zoological Survey of India* 84, 1–111.

Karny H, Docters van Leeuwen W, and Docters van Leeuwen-Reijnvaan J 1913. Beiträge zur Kenntnis der Gallen von Java. 5. Über die Javanischen Thysanopterocecidien und deren Bewohner. *Bulletin du Jardin Botanique de Buitonzorg* 10, 1–54.

Korneyev V, Zwölfer H, and Seitz A 2005. Phylogenetic relationships, ecology, and ecological genetics of cecidogenous Tephritidae, pages 321–372. *In* Raman A, Schaefer CW, & Withers M (Eds.) *Biology, ecology, and evolution of gall-inducing arthropods.* Science Publishers, New Hampshire, USA. pages 321–372.

Krishnamurthy KV, Raman A, and Ananthakrishnan TN 1975. On the morphology of the ceratoneon thrips galls of *Schefflera racemosa* Harms. *Marcellia* 38, 179–184.

Labandeira CC 1997. Insect mouthparts: ascertaining the paleobiology of insect feeding strategies. *Annual Review of Entomology* 28, 153–193.

Labandeira CC 2007. The origin of herbivory on land: initial patterns of plant tissue consumption by arthropods. *Insect Science*14, 259–275.

La Salle J 2005. Biology of gall inducers and evolution of gall induction in Chalcidoidea (Hymenoptera: Eulophidae, Eurytomidae, Pteromalidae, Tanaostigmatidae, Torymidae), *In* Raman A., Schaefer CW, & Withers TM (Eds.) *Biology, ecology, and evolution of gall-inducing arthropods.* Science Publishers, New Hampshire, USA. pages 539–572.

Lushai G, Markovitch O, and Loxdale HD 2002. Host-based genotype variation in insects revisited. *Bulletin of Entomological Research* 92, 159–164.

Mani MS 1959. Cecidotheca Indica. *Agra University Journal of Research (Science)* 8, 91–220.

Mani MS 1964. *The ecology of plant galls.* Walter Junk Publishers, The Hague. The Netherlands. .

Mani MS 1973. *Plant galls of India* (First edition). Macmillan India, New Delhi, India.

Mani MS 2000. *Plant galls of India* (Second edition). Science Publishers, New Hampshire, USA.

Martinez J-JI, Mokady O, and Wool D 2005. Patch size and patch quality of gall-inducing aphids in a mosaic landscape in Israel. *Landscape Ecology* 20, 1013–1024.

Mathur RN 1935. Notes on the biology of Psyllidae. *Indian Forest Records* 1, 35–70.

Mathur RN 1946. On the immature stages of some Psyllidae. *Indian Journal of Entomology* 8, 224–236.

Mathur RN 1975. *The Psyllidae of the Indian subcontinent.* The Indian Council of Agricultural Research New Delhi, India.

Mayr E 1963. *Animal species and evolution.* Harvard University Press, Cambridge, Massachusetts. USA.

Mendel, Z, Protasov, A, Fisher N, and La Salle, J 2004. Taxonomy and biology of *Leptocybe invasa* gen. & sp. n. (Hymenoptera: Eulophidae), and invasive gall inducer on *Eucalyptus*. Australian Journal of Entomology 43, 101–113.

Mendonça MDS 2007. Plant diversity and galling arthropod diversity searching for taxonomic patterns in an animal—plant interaction in the Neotropics. *Boletín de la Sociedad Argentina de Botánica* 42, 347–357.

Meyer J 1987. *Plant galls and gall inducers.* Gebrüder Bornträger, Stuttgart, Germany.

Mitchell, JD 1990. The poisonous Anacardiaceae genera of the world. *Advances in Economic Botany* 8, 103–129.

Morris DC, Schwarz MP, Cooper SJB, and Mound LA 2002. Phylogenetics of Australian *Acacia* thrips: the evolution of behavior and ecology. *Molecular Phylogenetics and Evolution* 25, 278–292.

Mound LA 1994. Thrips and gall induction: a search for patterns. *In* Williams MAJ (Ed.) *Plant galls: organisms, interactions, populations.* Systematic Association Special Volume 49, Clarendon Press, Oxford, UK. pages 131–149.

Mound LA and Kranz BD 1997. Thysanoptera and plant galls: towards a research programme. *In* Raman A (Editor) *Ecology and evolution of plant-feeding insects in natural and man-made environments.* Backhuys Publishers, Leiden, The Netherlands. Pages 11 –24.

Mound LA and Morris DC 2005. Gall-inducing thrips: an evolutionary perspective. *In*: Raman A, Schaefer CW, & Withers TM (Editors) *Biology, ecology, and evolution of gall-inducing arthropods.* Science Publishers, New Hampshire, USA. pages 59–72.

Muller J 1984. Significance of fossil pollen for angiosperm history. *Annals of the Missouri Botanical Garden* 71, 419–443.

Mukherjee SK 1997. Introduction: botany and importance. *In*: Litz RE (Ed.) *The mango – botany, production and uses.* CAB International, Oxford, UK. pages 1–20.

Nott PE and Roberts JC 1967. The structure of mangiferin. *Phytochemistry* 6, 741–747.

Pandit SS, Mitra S, Giri AP, Pujari KH, Patil BP, Jambhale ND, and Gupta VS 2007. Genetic diversity analysis of mango cultivars using inter simple sequence repeat markers. *Current Science* 93, 1135–1141.

Pell SK 2004. *Molecular Systematics of the Cashew Family (Anacardiaceae).* PhD dissertation. Louisiana State University, Baton Rouge, Louisiana, USA.

Raman A 1987. On the cecidogenesis and nutritive tissues of the leaf galls of *Garuga pinnata* Roxburgh (Burseraceae) induced by *Phacopteron lentiginosum* Buckton (Pauropsyllinae, Psyllidae, Homoptera). *Phytophaga* 1, 141–160.

Raman A 1991. Cecidogenesis of leaf galls on *Syzigium cumini* (L.) Skeels (Myrtaceae) induced by *Trioza jambolanae* Crawford (Homoptera: Psylloidea). *Journal of Natural History* 25, 653–663.

Raman A 1996. Nutritional diversity in gall-inducing insects and their evolutionary relationships with flowering plants. *International Journal of Ecology & Environmental Sciences* 22, 150–160.

Raman A 2007a. Insect-induced plant galls of India: unresolved questions. *Current Science* 92, 748–757.

Raman A 2007b. Biogeographical implications in species richness, biological diversity, and evolution of gall-inducing insects of the Orient and the eastern Palearctic. *Oriental Insects* 41, 9–25.

Raman A and Ananthakrishnan TN 1983. On the developmental morphology of the rosette galls of *Acacia leucophloea* Willd. (Mimosaceae) induced by *Thilakothrips babuli* Ramk. (Thysanopetra: Insecta). *Proceedings of the Indian Academy of Sciences* 92, 343–350.

Raman A and Ananthakrishnan TN 1985. Cecidogenous *Crotonothrips* (Thysanoptera) and *Memecylon* interactions: Host relations, nutritive tissue, tissue dynamics, and cecidogenetic patterns. *Proceedings of the Indian Academy of Sciences* 95, 103–120.

Raman A and Takagi S 1992. Galls induced on *Hopea ponga* (Dipterocarpaceae) in southern India and their gall-maker belonging to the Beesoniidae. *Insecta Matsumurana (New Series)* 47, 1–32.

Raman A and Withers T 2003. Oviposition by introduced *Ophelimus eucalypti*. (Hymenoptera: Eulophidae) and morphogenesis of female-induced galls on *Eucalyptus saligna* (Myrtaceae) in New Zealand. *Bulletin of Entomological Research* 93, 55–63.

Raman A, Singh RN and Maryanska–Nadachowska A 1996. Biology and karyology of a cecidogenous psylloid, *Trioza fletcheri minor* (Homoptera: Psylloidea) and morphogenesis of galls on the leaves of *Terminalia tomentosa* and *Terminalia arjuna*. *Insecta Matsumurana (New Series)* 53, 117–134.

Raman A, Schaefer CW, and Withers TM 2005. Galls and gall-inducing arthropods: An overview of their biology, ecology, and evolution. *In* Raman A, Schaefer CW, & Withers TM (Editros) *Biology, ecology, and evolution of gall-inducing arthropods*, Science Publishers, Inc., Enfield, New Hampshire, USA. pages 1–33.

Raman A, Schaefer CW, and Withers CM (Eds.) 2005b. *Biology, ecology, and evolution of gall-inducing arthropods*. Science Publishers, New Hampshire, USA.

Raman A, Cruz ZT, Muniappan R, and Reddy GVP 2007. Biology and host specificity of gall-inducing *Acythopeus burkhartorum* (Coleoptera: Curculionidae), a biological-control agent for the invasive weed *Coccinia grandis* (Cucurbitaceae) in Guam and Saipan. *Tijdschrift voor Entomologie* 150, 181–191

Raman A, Burckhardt D, and Harris KM 2009. Biology and adaptive radiation in the gall-inducing Cecidomyiidae (Insecta Diptera) and Calophyidae (Insecta Hemiptera) on *Mangifera indica* (Anacardiaceae) in the Indian subcontinent. *Tropical Zoology* 22, 000–000 (in press).

Rohfritsch O 1971. Étude d'une galle de *Lobopteromyia* sp. sur *Acacia* (*A. ferruginea* DC?). *Marcellia* 37, 139–149.

Rohfritsch O 2008. Plants, gall midges, and fungi: a three-component system. *Entomologia Experimentalis et Applicata* 128, 208– 216.

Rokas A, Melika G, Abe JL, Nives—Aldrey JL, Cook JM, and Stone GN 2003. A mitochondrial molecular phylogeny of the European gall wasps (Hymenoptera: Cynipidae: Cynipini). *Molecular Phylogeny & Evolution* 2, 36–45.

Roskam HC 2005. Phylogeny of gall midges. In: Raman A, Schaefer CW, & Withers TM (Editors) *Biology, ecology, and evolution of gall-inducing arthropods.* Science Publishers, New Hampshire, USA. pages 305–319.

Saxena G, McCutcheon AR, Farmer S, Towers GHN, and Hancock REW 1994. Antimicrobial constituents of *Rhus glabra. Journal of Ethnopharmacology* 42, 95–99.

Schaefer CW, Raman A, and Withers TM 2005. Galls and gall-inducing arthropods; ecological issues and evolutionary patterns. *In* Raman A, Schaefer CW, & Withers CM (Editors) *Biology, ecology, and evolution of gall-inducing arthropods.* Science Publishers. New Hampshire, USA. pages 761–766.

Schoonhoven LM, van Loon JJA, and Dicke M 2005. Insect-plant biology. Oxford University Press, Oxford, UK.

Scott AC, Stephenson J, and Collinson ME 1994. The fossil record of leaves with galls. In Williams MAJ (Editor), *Plant gall – organisms, interactions, and populations,* The Systematics Association Special Volume 49, Clarendon Press, Oxford, UK. pages 447–470.

Shorthouse JD, Wool D, and Raman A 2005. Gall-inducing insects—nature's most sophisticated herbivores. *Basic and Applied Ecology* 6, 407–411.

Sokhi J and Kapil RN 1984. Morphogenetic changes induced by *Trioza* in flowers of *Terminalia arjuna* – I. Androecium. *Phytomorphology* 34, 117–128.

Sokhi J and Kapil RN 1985. Morphogenetic changes induced by *Trioza* in flowers of *Terminalia arjuna* –II. Gynoecium. *Phytomorphology* 35, 69–82.

Srivastava AK, Abbas SR, Mehrotra RC, and Srivastava R 2000. Cecidomyiid leaf galls in Palaeocene leaves from north-eastern India. *Acta Palaeobotanica* 40, 43–47.

Stireman JO (III), Nason JD, Heard SB, and Seehawer JM 2006. Cascading host-associated genetic differentiation in parasitoids of phytophagous insects. *Proccedings of the Royal Society (Biological Sciences)* 273, 523–530.

Stone GN and Schönrogge K 2003. The adaptive significance of insect gall morphology. *Trends in Ecology & Evolution* 18, 512–522.

Stone GN, Atkinson R, Rokas A, Csóka G. and Nieves-Aldrey J.L. 2001. Differential success in northwards range expansion between ecotypes of the marble gallwasp *Andricus kollari*: a tale of two lifecycles. *Molecular Ecology* 10, 761–778.#

Strong DR, Lawton JH, and Southwood TRE 1984. Insects on plants—community patterns and mechanisms. Blackwell, Oxford, UK.

Sýlven E 1979. Gall midges (Diptera, Cecidomyiidae) as plant taxonomists. *Acta Universitatis Upsaliensis* 22, 63–69.

Takagi S 1987. Notes on the Beesoniidae (Homoptera: Coccoidea). *Insecta Matsumurana (New Series)* 37, 27–41.

Takagi S 1995. A new species of *Beesonia* with larval polymorphism, inducing a stem gall on *Shorea curtisii* in Singapore (Homoptera: Coccoidea: Beesoniidae). *Insecta Matsumurana (New Series)* 52, 1–19.

Takagi S 2001. Four gall-inducing eriococcids, with notes on dipterocarp-associated gall-inhabiting coccoids (Homoptera: Coccoidea: Eriococcidae and Beesoniidae). *Insecta Matsumurana (New Series)* 58, 51–113.

Takagi S 2007. The gall-inducing coccoid family Beesoniidae (Hemiptera): facts, speculations, and perspectives. *Oriental Insects* 41, 67–91.

Thomas CD and Singer MC 1998. Scale-dependent evolution of specialization in a checkerspot butterfly: from individuals to metapopulations and ecotypes. *In* Mopper S & Strauss SY (Editors) *Genetic structure and local adaptation in natural insect populations*. Chapman & Hall, London, UK. pages 343–374.

Thompson 1994 Thompson, J.N. 1994. *The coevolutionary process*. The University of Chicago Press, Chicago, USA.

Tokuda M, Harris KM, and Yukawa J 2005. Morphological features and molecular phylogeny of *Placochela* Rübsaamen (Diptera: Cecidomyiidae) with implications for taxonomy and host specificity. *Entomological Science* 8, 419–427.

Uechi N, Kawamura F, Tokuda M, and Yukawa J 2002. A mango pest, *Procontarinia mangicola* (Shi) comb. Nov. (Diptera: Cecidomyiidae), recently found in Okinawa, Japan. *Applied Entomology & Zoology* 37, 589–593.

Uechi N, Yukawa J, and Yamaguchi D 2004. Host alteration by gall midges of the genus *Asphondylia* (Diptera: Cecidomyiidae). *Bishop Museum Bulletin of Entomology* 12, 53–66.

Uichanco HR 1919. A biological and systematic study of Philippine plant galls. *Philippine Journal of Science* 14, 527–554.

Varadarasan S and Ananthakrishnan T N 1982. Biological studies on some gall thrips. *Proceedings of the Indian National Science Academy Part B* 48, 35–43.

Wool D 2005. Gall-inducing aphids: biology, ecology, and evolution; in Biology, ecology, and evolution of gall-inducing arthropods. *In* Raman A, Scahefer CW, and Withers TM (Eds.) *Biology, ecology, and evolution of gall-inducing arthropods*. Science Publishers New Hampshire, USA. pages 73–132.

Yang MM, Tung GS, La Salle J, and Wu ML 2004. Outbreak of erythrina gall wasp on *Erythrina* spp. (Fabaceae) in Taiwan. *Plant Protection Bulletin* 46, 391–396.

Yang MM and Raman A 2007. Diversity, richness, and patterns of radiation among gall-inducing psyllids (Hemiptera: Psylloidea) in the Orient and Eastern Palearctic. *Oriental Insects* 41, 55–65.

Yang S-Y, Chen M-Y, and Yang J-T 2002. Application of cecidomyiid galls to the systematic of the genus *Machilus* (Lauraceae) in Taiwan. *Botanical Bulletin, Academia Sinica* 43, 31–35.

Yukawa J and Masuda H 1996. *Insect and mite galls of Japan in colors.* Zenkoku Nôson Kyôiku Kyôkai, Tokyo, Japan.

Yukawa J and Partomihardjo O 1997. Insect and mite galls collected from Peucang, Panaitan, and Krakatau islands, Indonesia. *Tropics* 7, 141–152.

Yukawa J, Uechi N, Horikiri M, and Tuda M 2003. Description of the soybean pod gallmidge, *Asphondylia yushimai* sp.n. (Diptera: Cecidomyiidae), a major pest of soybean and findings of host alteration. *Bulletin of Entomological Research* 93, 73–86.

Yukawa J and Rohfristch O 2005. Biology and ecology of gall-inducing Cecidomyiidae (Diptera) In: Raman A, Schaefer CW, & Withers TM (Editors) *Biology, ecology, and evolution of gall-inducing arthropods.* Science Publishers, New Hampshire, USA. pages 273–304.

Yukawa J, Uechi N, Tokuda M and Sato S 2005. Radiation of gall midges (Diptera: Cecidomyiidae) in Japan. *Basic and Applied Ecology* 5, 453–461.

Chapter - 10

Biodiversity and Ecosystem level consequences in relation to Thrips - host interactions

T.N. Ananthakrishnan

Flat 6, Dwaraka, 22, Kamdarnagar Ist Street, Chennai- 600 034

Introduction

The gap between population and ecosystem ecology has narrowed considerably with their interlinking having come to be recognized. The classifications of species into functional groups appear as a way to simplify assessment of species effect on ecosystems. The composition of biological communities varies greatly from place to place and from one season to the next. Functional groups need to be defined and accordingly species are now grouped based on morphological, physiological, and phenological attributes. Studies on the effects of diversity within functional groups will depend on the precision of the interactions involved. In the absence of extensive field studies, biodiversity studies cannot provide adequate results, to be able to confidently point out species level identifications, for which an understanding of structural and biological variations become vital. Intraspecific polymorphism therefore, is often indicated as an operational problem for species recognition for their connection. Mound (2002) states, ``Taxonomy contributes directly to the ultimate objective of all biological studies towards understanding the origin and maintenance of biological diversity. Viewed in this way, Taxonomy and systematics have much broader objectives than the mere description and cataloguing of species. To achieve such an objective, taxonomic studies must involve good field biology, in order to recognize structural and biological variations". It is with this objective that this exercise seeks to enumerate the bioecological perspective of thrips, emphazing aspects such as biological diversity, bioecology and behaviour. Determined effort to identify diverse niches of thrips from different plant formations in diverse ecoregions such as Western and Eastern Ghats, the Aravalli raanges, foothills of Himalayas, besides the desert and semidesert regions, undoubtedly resulted in the recognition of several species. From the distributional view point, the Indian thrips fauna involves the Oriental, Ethiopian, Palaearctic and Nearctic elementsd and the need to discover more endemic species calls for a better assessment of taxonomic aspects.

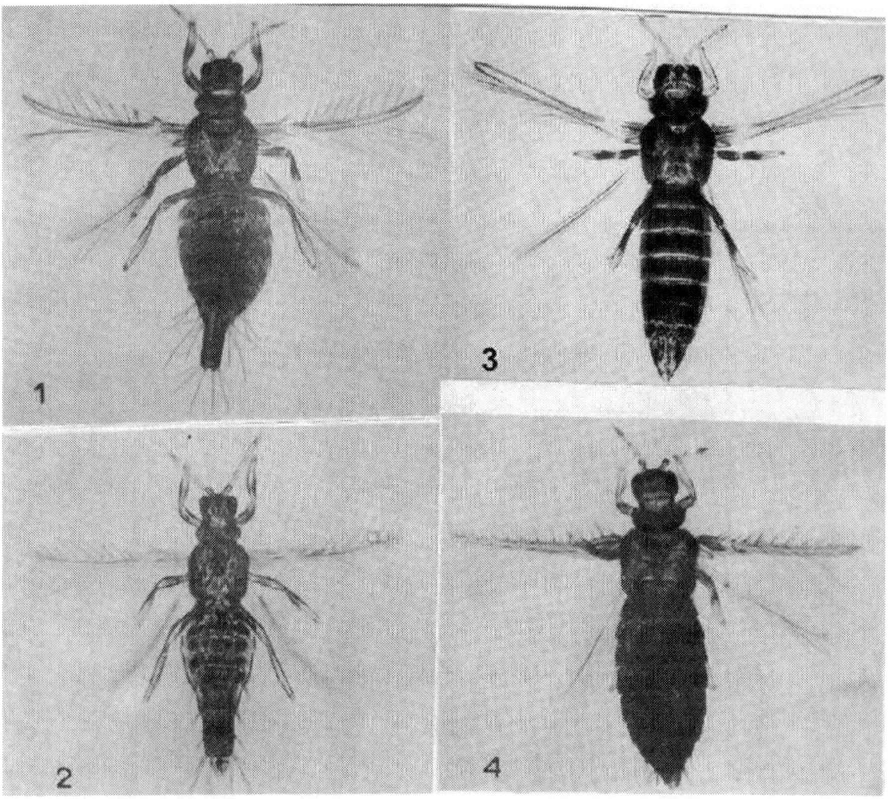

Fig. 1. *Panchaetothrips indicus* (Female) (x45); (2) *Panchaetothrips indicus* (male) (x30); (3) *Selenothrips rubrocinctus* (female) (x45); and (4) *Heliothrips haemorrhoidalis* (female) (x45)

The composition of biological communities varies from place to place and from one season to the next. Their adaptive diversity has enabled successful exploitation of diverse niches in a variety of plant formations, as well as fungus-infested habitats such as plant litter and bar of living and dead trees. The increasing evidence of thrips infestation in various cropping systems, their ability to migrate from weed reservoirs to the crops and vice-versa, and their intercrop movement and related adaptive strategies are basic aspects underlying the bionomics of thrips. Indepth studies on the nature of population increase in different agro-ecosystems such as cereals and millets, pulse crops, fibre and plantation crops indicate the need to consider crops as a unit integrated with the adjacent vegetation. Factors such as number of species, abundance, intra and their habitat diversity, resource partitioning are important in structuring thrips communities. Their greater the number of effective habitats within an ecosystem, greater is the diversity of thrips. Grasslands, field crops, forests and closed gall communities comprise specific microhabitats for thrips. As such the abundance and patterns of distribution of various host plants involving crops as well as alternate weed hosts are important factor in the population dynamics of thrips. Indepth studies on the nature of population increases in different agroecosystems viz. cereals, millets, pulse crops, fibre plantation crops, oil seed crops are basic to our understanding of the bionomics of thrips. Factors such as number of species, abundance, intra and inter habitat diversity, resource partitioning are important in structuring thrips communities. The greater the number of effective habitats within an ecosystem, greater is the diversity of thrips. The fact cannot be overlooked, that throughout the season, the composition of the crop changes enabling movement of species within the plant niches. Besides thrips-host-weed interactions, the increased role of thrips in pollination, colour patterns and fragrance of flowers and quality of nectar, have considerable influence on host location and colonisation. An interesting aspect is their ability to adapt to various environments, their abundance being correlated with particular types of plant formation. Many Terebrantia show vertical stratification with different species occupying different nodes, tending towards avoiding competition. Many others form galls and yet others are mycophagous or fungus feeders, besides predatory species feeding on aphids, coccids, mites and thrips. Some are essential elements of the edaphon occurring as pupae and pre-pupae upto a depth of 10-30 cms in the soil. Dead and decaying vegetation, barks of trees and litter harbour several mycophagous species. The abundance and patterns of distribution of various hosts involving crops as well as weed hosts are important factors in the population dynamics of thrips. Indepth studies on the nature of population increases in different agroecosystems indicate the need for considering crops as a unit integrated with the adjacent vegetation. Needless to emphasize that factors such as number species, abundance, their intra and interhabitat diversity and overlap are important in structuring thrips communities, the complexity of which is enhanced with the inclusion of factors such as altitude, moisture and temperature governing the microclimate. The spatial and temporal variation have fascinated ecologists, with biodiversity attempting to describe the variety of living insects and the ecological complexes of which they are a part. Hence, taking thrips communities, an attempt is made here to assess the implication of diverse hosts to diverse thrips species, involving a taxonomic and ecological perspective.

Fig. 2. 1,2 *Rhipiphorothrips cruentatus* (left, female, right male); 3. *Retithrips syriacus* left, female; right male.

Grass infesting thrips

Four distinct typs of thrips communities are recognizable; phyllophilous, anthophilous, bark and litter communities and gall communities. The principal types of phyllophilous communities - grassland, crop, forest harbour their own representative communities; theire exist several microhabitats very often influenced by local differences in microclimate as evident within grass sheath. For a proper visualization of community structure in cereal crops, consideration of the total environment involving crops, grasses and numerous weed species appears a basic necessity. Thrips communities in such crops as rice, wheat, Sorghum result from the aggregation of species associated with related plants. Besides continually migrating to the crops these weeds serve as reservoirs of thrips communities. It must be remembered that there are more than twice as many species in the grasslands and weeds. There is a greater specific variety in grasslands with better numerical balance between its species. Studies on the population of thrips infecting graminaceous and cyperaceous weeds in paddy, sorghum etc have indicated that the dominant species in paddy is *Haplothrips ganglbaueri, Florithrips traegardhi* in Sorghum and *Chirothrips mexicanus* and *Exothrips hemavarna* in Pennisetum. The thrips community in the same crop may have different composition, ecological succession and dominant species. The graminaceous weed *Echinochloa crusgalli* harbours large populations of *Haplothrips ganglbaueri*. Changes in the patterns of community structure in thrips are often noticeable and the more common grass inhabitants are —

While it is natural to classify grass infesting thrips as leaf, sheath, leaf blade and inflorescence inhabitants the degree of infestation is of importance, particularly when several grasses harbour a number of thrips species, some primary inhabitants, other secondary and yet other casual or rare. The more common grass inhabitants are —

Biodiversity of grass infesting thrips

Andropogon pertusus	*Anaphothrips flavicinctus*
	Neolimothrips saccharina
	Sericothrips graminis
	Phibalothrips peringueyi
	Podathrips oryzae
	Hoplandrothrips indicus
	Haplothrips apicalis
	Haplothrips gowdeyii
Apluda aristata	*Anaphothrips sakimurai*
	Apinothrips rufus
	Caliothrips graminicola
	Caliothrips indicus
	Chirothrips manicatus
Chloris barbata	*Neolimothrips saccharivora*
	Chiraplothrips priesneri

	Phibalothrips peringueyi
Cynodon dactylon	*Haplothrips apicalis*
Cymbapogon citratus	*Praepodothrips cymbapogonii*
Eragrostis Sp.	*Caprithrips analis*
	Aptinothrips rufus
Oryza sativa	*Thrips oryzae*
Panicum maximum	*Anaphothrips flavicinctus*
	Chirothrips maximi
	Exothrips madrasensis
	Caliothrips indicus

Fig. 3. Top (left) *Strepterothrips orientalis*, Oedymerous male (apterous), (right) *Idiothrips fici*, Female. Bottom (left) *Strepterothrips orientalis*, macropterous female; (right) *Stictothrips fimbriata* Female.

Flower infesting thrips

Anthophilous Communities

Aggregation of heavy thrips population is generally met within the flowers and inflorescence of several plants, the aggregation behaviour being perhaps to ensure that the sexes are brought together. The number of flowers in an area fluctuates regularly with seasons, long and dry summers producing very few flowers, with profuse flowering in spring, decreasing in winter, so that seasonal changes with access to food appear related to changes in the abundance of thrips communities. The ecological succession of thrips communities on *Ruellia tuberosa* in Chennai revealed the relative duration of different thrips species within flowers leading to their recognition as primary, secondary and tertiary inhabitants. *Haplothrips gowdeyi* and *Frankliniella schultzei* appear first and occur thoroughout the flowering season, are the primary species. *Megalurothrips distalis* and *Thrips hawaiiensis* are secondary inhabitants displaying irregularity in their population trends. Tertiary species occur in lesser number and occur only for a short period of time. Thrips communities within inflorescence of Compositae reveal the seasonal incidence of host plants as well as thrips inhabiting them.

Even as grasses form sizable hosts for several species of thrips, flowers are inflested by an equal number of species, the more common ones being —

Thrips palmi

Thrips flavus

Thrips hawaiiensis

Microcephalothrips abdominalis

Frankliniella schultzei

Haplothrips ganglbaueri

Megalurothrips distalis

Sciothrips cardamomi

Astrothrips tumiceps

Isothrips orientalis

Haplothrips tardus

Tusothrips aureus

Frankliniella intonsa

Vertical stratification with different species inhabiting different nodes, thereby avoiding competition is common. Many others form galls and yet others are mycophagous or fungus feeders, besides predatory species feeding on aphids, coccids, mites and thrips. Some are essential elements of the edaphon occurring as pupae or prepupae upto a depth of 10-30 cm in the soil. Dead and decaying vegetation, barks of trees and litter harbour several mycophagous species.

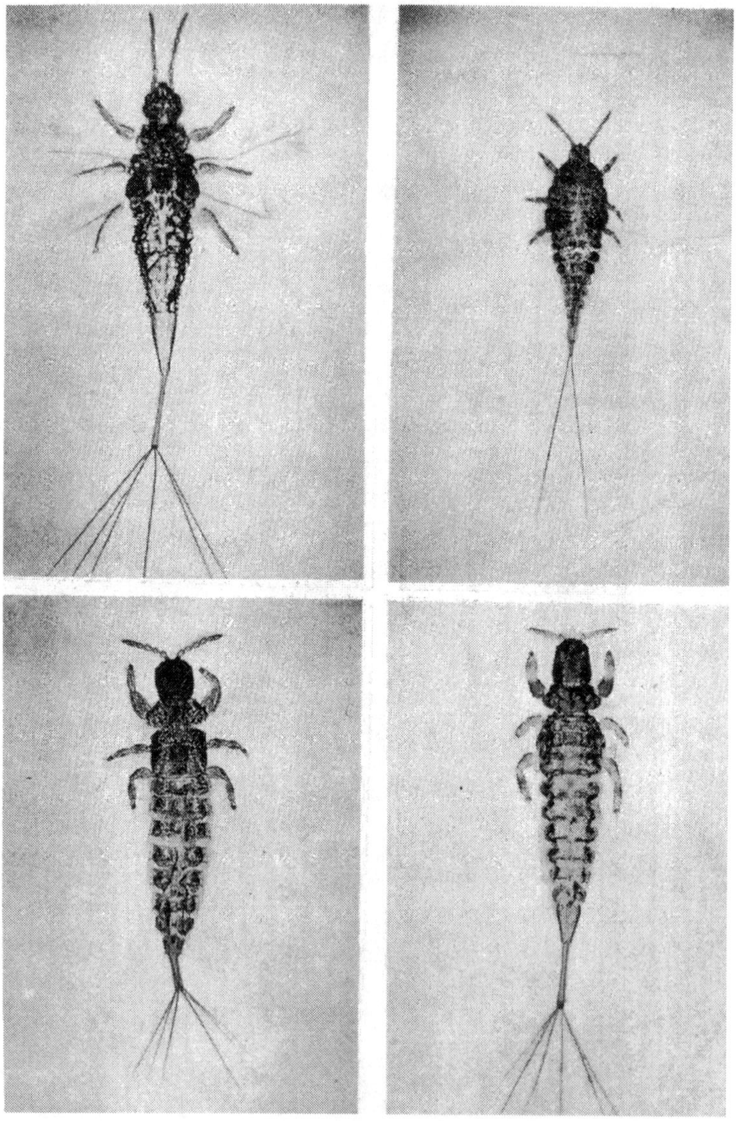

Fig. 4. Urothripids

Mycophagous Thrips

Several species of mycophagous thrips occur in the bark or ded branches of trees and appear to be dependent on fungal hosts for food, shelter and a breeding place. These are the mycobionts as aparts from the sporobionts totally dependent on fungal spores. Further, the ability of many of these to effectively reproduceby oviparity, ovovivi-and viviparity appears an important factor. Several species have developed the tendency to aggregate, the nature of the aggregation and population size depending on fungal food availability. Species of *Haplothrips* form large aggregates, feeding directly in fungal aggregates: utilization of diverse ecological which is very typical of sporophagous Tubulifera, and such categories are (a) species exploiting various host for fungal food (b) species mostly host specific, host feeding on a wide variety of fungi present, such as Ascomycetes, coelomomycetes and hyphomycetes (c) species occurring in a wide range of hosts, mostly drying grain clumps. Sex-limited polymorphism is very typical of the males, resulting in oedy-merism and gynaecoidism and major, normal and minor categories among females. The type of reproduction - whether ovi-vivi- or ovoviviparity is determined by factors such as enviro-nment, fungal food availability, aggregations. Viviparity and ovoviviparity are adaptation to shorten the life cycle enabling young ones to complete development before food disappears.

Association of thrips with plant pathogenic fungi is an aspect demanding increased attention. The dual role played by many of the thrips as mechanical vectors as well as efficient spore feeders of plant pathogenic fungi becomes relevant. In view of the large number of fungal spores consumed by an individual at a given time, the overall consumption of a large quantity of inoculum by a number of aggregating population at a given area during the whole life cycle becomes important. This function of consumption and dissemination of pathogenic fungal spores appear essential aspects of the biology of sporophagous thrips. Of several species recorded by the author over the past many years, one could confidently look for abundant material such as species of *Nesothrips, Kleothrips, gigans, Elaphrothrips productus, E. beesoni, Hoplandrothrips flavipes, Azaleothrips amabilis, Strepterothrips orientalis, Polyphemothrips cracens, Hoplothrips fungosus, Ecacanthothrips sanguineus, Dinothrips sumatrensis, Diceratothrips usitatus,* to mention the more common ones. The mycelial feeders consume *Aspergillus niger,*

Ganoderma, Pestalotia, Dreschlera, while the sporophagus feeders consume spores of *Aspergillus niger, Ganoderma, Pestalotia, Phomopsis tectonae,* to mention the more important. As such feeding diversity is respect of fungal resources utilization in diverse ecological niches is very typical of sporophagous Idolothripines, enabling recognition of the following distinct categories —

(a) Species which exploits hosts for fungal resources, but feed only on the spores of the same fungus as in *Dinothrips sumatrensis* feeding only on the spores of *Lasiodiplodia.*

(b) Species which are host specific, but feed on a wide variety of fungi present, involving all fungi as in *Tiarothrips subramanii* feeding on *Anthostomella corsanguinea, A. sepilibilis, A. phoenicicola, Pestalozzia* and *Melanographium citri.*

(c) Species with a restricted host range and restricted feeding as in *Elaphrothrips denticollis* occurring in large numbers in drying leaves of *Areca catechu, Tectona grandis* feeding on the spores of *Pestalozzia* and *Phomopsis; Bactrothrips idolomorphus* on dry leaves of *Shorea robusta* feeding on *Pestalozzia* and *Lasiodiplodium.*

(d) Species occurring on a wide range of hosts, mostly drying grass clumps.

Diversity of mycophagous thrips

Around 60 genera including about 135 species of which around 80 are Phlaeothripines and the next Megathripines.

Phlaeothripinae		Megathripinae	
Ablemothrips maxillatus	Anan.	*Allothrips indicus*	Anan.
Adraneothrips madrasensis	Anan.	*Allothrips bicolor*	Anan.
Adraneothrips disjinctus	Anan.	*Allothrips montanus*	Anan.
Adraneothrips stannardis	Anan.	*Bactrothrips idolomorphus*	Karny
Adraneothrips elegans	Anan.	*Bactrothrips luteus*	Anan.
Agrothrips insolitus	Anan.	*Decothrips anacardii*	Anan.
Azaleothrips amaloilis	Anan.	*Diaphorothrips thevetii*	Anan.
Bunothrips cruralis	Anan.	*Diceratothrips usitatus*	Anan. & Jag.
Callothrips ascitus	Anan.	*Diceratothrips brevisetosus*	Anan. & Jag.
Ecacanthothrips sanguineus	Bagnall	*Diceratothrips niger*	Bagnall
Hoplandrothrips graminis	Anan.	*Diceratothrips uredinus*	Anan. & Jag.

Hoplandrothrips corticis	Anan.	*Diceratothrips indicus*	Bagnall
Hoplandrothrips flavipes	Bagnall	*Diceratothrips vitreipennis*	Priener
Hoplothrips fungosus	Moulton	*Diceratothrips longisetosus*	Anan. & Jag.
Hoplothrips nemorius	Anan.	*Dinothrips sumatrensis*	Bagnall
Hoplothrips orientalis	Anan.	*Elaphridia crassipes*	Bagnall
Idiothrips ficus	Bhatti	*Elaphrothrips productus*	Procer Schmutz
Karnyothrips flavipes	Jones	*Elaphrothrips beesoni*	Bagnall
Liophlaeothrips nitidus	Anan.	*Elaphrothrips nicidus*	Anan.
Liophlaeothrips pictus	Anan.	*Elasphrothrips notabilis*	Anan.
Macrophthalmothrips splendidus	Anan.	*Elaphrothrips secus*	Anan.
Malacothrips lucidus	Anan.	*Elaphrothrips clarispinis*	Anan.
Malacothrips madrasensis	Anan.	*Elaphrothrips procer*	Schmutz
Malacothrips indicus	Anan.	*Kleothrips gigans*	Bagnall
Margaritothrips sumatrensis	Priesner	*Loyolaia indica*	Anan.
Margaritothrips flavus	Anan.	*Machatothrips biuncinatus*	Anan.
Margaritothrips longus	Anan.	*Machatothrips indicus*	Anan.
Mystrothrips dammermani	Priesner	*Machatothrips silvaticus*	Anan.
Nesothrips lepidus	Anan.	*Machatothips corticosus*	Anan.
Neurothrips indicus	Anan.	*Meiothrips menoui*	Anan.
Phlaeothrips nilgiricus	Anan.	*Nesothrips karnyi*	Anan.
Phylladothrips karnyi	Priesner	*Nesothrips indicus*	Anan.
Plectrothrips corticinus	Anan.	*Nesothrips alius*	Anan.
Plectrothrips pallipes	Anan.	*Nesothrips robustus*	Anan.
Polyphemothrips cracens	Anan.	*Nesothrips falcatus*	Anan.
Polyphemothrips enormis	Anan.	*Nesothrips acuticornis*	Anan.
Polyphemothrips megacephalus	Anan.	*Oedaleothrips congoensis*	Hood
Polyphemothrips mirandus	Anan.	*Oedaleothrips ramamurthii*	Anan.
Polyphemothrips fumidus	Anan.	*Paxillothrips longicaudus*	Anan.
Polyphemothrips ruidus	Anan.	*Percnothrips turbinatus*	Anan.
Polyphemothrips stannardi	Anan.	*Phoxothrips faurei*	Anan.
Socothrips verrucosus	Anan.	*Priesneriana kabandha*	Ramakrishna
		Pygothrips amplus	Anan.
		Rhaebothrips lativentris	Karny
		Stinothrips typicus	Anan.
		Tiarothrips subramanii	Ramakrishna
		Uredothrips indicus	Anan.
		Uredothrips tirumalaiensis	Anan.

Association of thrips with plant pathogenic fungi is an aspect demanding increased attention, both in view of the number of species associated, as well as the number individuals comprising them. Association such as *Megaphysothrips subramanii* with *Hamelia vastatrix* on coffee, *Chaetanaphothrips signipennis* with the banana peel rust, *Stigmothrips fimbriata* with *Rhizopus stolonifera* are a few examples.

Fig. 5. *Left* Ficus galls, *right* Pouch gall of *Casearia tomentosa*

Gall forming Thrips

Thrips are frequently associated with structural abnormalities of active plant tissue termed galls or cecidia and cecidology today has become an important field of study. Galls are manifestations of growth abnormalities induced on a plant by insect and plants respond to the feeding stimulus by producing such abnormal growth. Galls may arise on roots, shoots, petioles, stipules, leaf blades, flower buds, inflorescence axis, bracts and flower. Though the tendency to produce galls is common, true galls are always positive directional responses and cecidogenesis involves the recanalization and reorientation of plant development resulting in the insects being partially or completely enclosed, so that the gall insects grow, mature and reproduce within the gall. Insect galls can vary from simple open deformation to highly complex and organized, enclosed structure. About 50% of thrips galls are leaf rolls, 25-30% leaf folds, 10-12% leaf curls adn 8-

10% where gall tissues are markedly different from the host leaf. Gall types are leaf olds, marginal leaf rolls, marginal complete rolls, lamina complete rolls, pouches, horn galls, rosette galls, tubular galls, flower galls and stem galls (Ananthakrishnan, 1992).

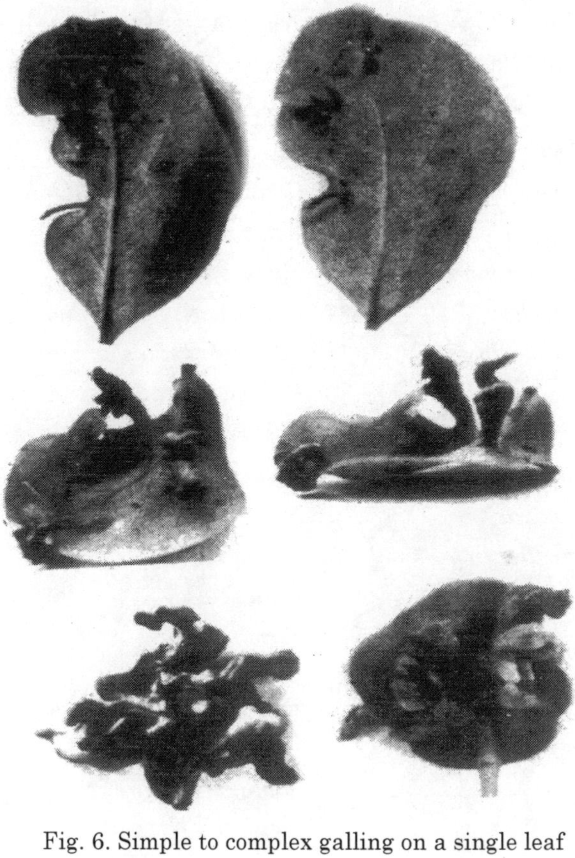

Fig. 6. Simple to complex galling on a single leaf

Shift of gall induction site is indicative of the adaptive capacity to evade seasonal non-availability of vegetative meristerms is evident in *Thilakothrips babuli* inducing rosette galls on shoot meristem of *Acacia leucophloea*. It also induces galls on floral meristems. Even as gall insects serve as reliable plants taxonomist, some gall bearing plants serve as useful `insect taxonomists' (Raman, 2007). This is exemplified by *Crotonothrips* and *Memecylon* very common in the Western and Western ghats., specifying the hosts *C. gallarum, C. coorgensis, C. dissimilis, C. memecylonensis* and *C. dantahasta* induce galls on different species of *Memecylon*.

Despite the apparent lack of any general pattern in host-exploitations several patterns can be detected at the generic and specific levels (Mound, 1994). Some of these are :- one thrips species/one plant species; one thrips species/one plant genus; several related thrips species/one plant species; several related thrips species/several related plant genera; several related thrips species/several congeneric plants; woody gall with thrips on casuarina trees.

Some of the more common hosts of gall thrips are :-

Diversity of gall thrips	Host
Aneurothrips punctipennis	*Cordia suavolens*
Alocothrips hadrocerus	*Gymnorporia*
Arrhenothrips ramakrishnae	*Mimusops elengi*
Austrothrips cochinchinensis	*Calycopteris floribundus*
Brachythrips dantahasta	*Memecylon edulc*
Cercothrips nigrodentatus	*Planchonia valida*
Dixothrips onerosus	*Casuarina tomentosa*
Gynaikothrips flaviantennatus	*Ficus* sp.
Gynaikothrips malabaricus	*Ficus* sp.
Gynaikothrips scheffléricola	*Piper nigrum*
Gynaikothrips viticola	*Ficus retusa*
Liothrips karnyi	*Terminalia* sp.
Liothrips malabaricus	*Linociera* sp.
Liothrips variabilis	*Mallotus* sp.
Liothrips tibialis	*Ficus benjamina*
Liothrips ramakrishnae	*Santalum album*
Liothrips uzeli	*Loranthus elasticus*
Lygothrips jambuvasi	*Walsura piscida*
Mesicothrips plicatus	*Ventilago* sp.
Mesothrips extensivus	*Pothos scandens*
Mesothrips jordani	*Acacia leucophloea*
Mesothrips manii	*Santalum album*
Phorinothrips loranthi	*Loranthus* sp.
Psenothrips priesneri	*Walsura piscida*
Schedothrips orientalis	*Ventilago maderaspatana*
Tetradothrips foliiperda	*Pothos scandens*
Thilakothrips babuli	*Acacia leucophloea*

Fig. 7. Some fungal hosts of thrips.

Vertical patterns of distribution

Ananthakrishnan *et al.* (1982) observed that the seven phytophagous thrips species (*Toxothrips ricinus, Scirtothrips dorsalis, Ayyaria chaetophora, Retithrips syriacus, Rhipiphorothrips cruentatus, Zaniothrips ricini* and *Astrothrips tumiceps*) on *Ricinus communis* exhibit a characteristic vertical distribution from the terminal bud downwards. *Toxothrips ricinus* is restricted to the buds covered by bracts and in stray cases, the larvae also occurred on the leaves of the first visible node from the terminal bud. *Scirtothrips dorsalis,* a pest of castor, was abundant both on the upper and lower surfaces of the tender leaves of the first and second nodes, rarely occurring on the third nodal leaves and inflorescence. The common polyphagous species, *R. syriacus,* a recognised pest of castor, mostly occurred from the leaves of third node downwards, but the population was comparatively more on both surfaces of the fourth and fifth nodal leaves. Along with *R. syriacus,* a meagre population of *R. cruentatus* and *A. tumiceps* was also present on both the surfaces of leaves from the third to seventh nodes. On the third and fourth nodal leaves, only females of *R. cruentatus* was noticed. The larvae and adults of *Z. ricini* were exclusively restricted to the abaxial surface of the leaves from the fourth node downwards up to the seventh node, and exhibited the same sex-ratio on all the leaves where they occurred. The infestation of *A. chaetophora* was evident from the fourth node onwards, with the rare incidence of adults on the tender leaves along the apical region.

In the case of *Arachis hypogaea* (Ananthakrishnan *et al.,* 1982) the young leaves of the first three nodes from the terminal were occupied by the polyphagous species *S. dorsalis* and rarely a few adults occurred in the flowers. Different life stages of this species including the eggs were collected from the leaves. The eggs were randomly distributed within the tissues of leaves of first three nodes and the females did not show preference to any leaf of a particular node for oviposition. The eggs were also noticed in the tender shoots. Raizada (1965) observed egg laying of this species inside any soft portion of the shoots on *Ricinus communis. C. indicus* occurred on the leaves of fourth node downwards and in stray cases, a few female adults were collected from second nodal leaves. Mostly the gravid females of *C. indicus* occurred on fourth nodal leaves, revealing their preference towards this region for egg laying. Two other species of Terebrantia, *Megalurothrips distalis* and *Frankliniella schultzei,* also infest *A. hypogaea,* but they are restricted to the flowers only and are totally absent during the non-flowering season.

On *Calotropis giganeta,* male and females of the polyphagous, anthophilous species *F. schultzei* occurred within the flowers and partially opened buds. In a few cases, adults were also noticed on the leaves of first visible node. But no larvae of this species were reported from this host plant. Leaves of the first three nodes, did not show any thrips infestation, but adults of *A. tumiceps* and *R. syriacus* were rarely noticed on the third nodal leaves. Rarely the third, and fourth nodal leaves were infested by *A. tumiceps* and *R. syriacus*. Both larvae and adults of the former species were observed in the leaves, but only the adults of *R. syriacus* were observed. Leaves of the fifth to seventh nodes exhibited a combined population of *R. syriacus* and *Z. ricini*. Sometimes the infestation of *Z. ricini* extended up to the leaves of eighth node, and below this (where old leaves were present) no thrips infestation was evident. Both adult and immature stages of *Z. ricini* occurred on the plant, and the females preferred the fifth nodal leaves. A mixed population of *A. tumiceps, R. syriacus* and *Z. ricini* infest the fifth and sixth nodal leaves. Thus, when each of these species preferred to inhabit a specific region, only the flowers and leaves of third to eighth nodes were infested. *F. schultzei* and *R. syriacus* visited the plants only for feeding, while *A. tumiceps* and *Z. ricini* utilised *Calotropis gigantea* for both feeding and reproduction.

Thrips on *Achyranthes aspera* also appeared to select a definite site of feeding, resting, and reproduction. These species exhibit a distinct vertical distributional pattern on the host plant. *Chiridothrips indicus* a tubuliferan (Ramakrishna and Margabandhu, 1932) was found to be a major species feeding on and breeding in the spikes of *A. aspera*. *F. schultzei,* the pollen-feeding anthopilous species occurred along the adaxial sides of the first nodal leaves of the plants bearing inflorescence, and fed mainly on the pollen grains shed from the flowers. Although no larvae could be collected from the first nodal leaves, adults were always found. Selection of the site by this species appears to be mainly for feeding and *F. schultzei* was completely absent during the non-flowering season of the plant. The second nodal leaves were generally occupied by *A. chaetophora* and *C. indicus* with both the sexes of former, but only the gravid females of the latter species. *C. indicus* thus appeared to prefer the second nodal leaves for egg-laying and when the second nodal leaves become third node, due to the production of a new pair of leaves, the gravid females migrated to the new second nodal leaves for egg-laying. Leaves of each node showed variation in the sex-ratio of the same thrips species. In plants which grow in dry areas and under big trees, the growth was seldom normal and the plants were comparatively smaller in size, and the vertical pattern of the thrips distribution was not very definite and the immature and adult stages were scattered over the plant.

Conclusion

Because of their abundance in phytophagus niches. thrips are associated with the types of plants formations and are known to be very susceptible to environmental change. They have a tendency to aggregate, forming small, discrete, but favourable microhabitats within the larger general habitats. As such phytophagus species are often designated as anthophilous, phyllophilous, and cecidicolous habitats. Adaptation of species with strong, well developed setae are perhaps designed to a better ability to carry pollen and the functional significance of these setae as pollen traps becomes important. The number of thrips pollinators also depend on the number and structure of setae as well as on the size and viscosity of pollen.

The fungus feeding habitat being widespread, has a wide impact on the biology of thrips, their dispersal activity and resultant speciation. The fluctuation in the habitat combined with internal environmental factor have possibly stimulated the evolution of winged and apterous morphs in many species as well as the production of structurally diverse forms resulting in oedymeres and gynaecoids males and major and minor females. Until one is fully aware of this spectrum of diversity, it could lead to considerable confusion regarding generic classificaton, not to mention of species determination. The lack of host specificity in mycophagous species is perhaps a basic reason for the complex relationships between various forms and with the fungal habitat being widespread, there is also wide impact on the biology of thrips, their dispersal activity and resultant speciation. Needless to emphasize the taxonomic richness and their distribution patterns are important aspects in studies pertaining to biodiversity. The more you delve deep into the habitats of diverse species, the more you are likely to strike a golden mine. This is very true of studies in biodiversity of thrips which offer on endless treasure if one looks into the intricacies of diverse habitats.

This review on the biodiversity of thrips in diverse environments would not have been possible, but for my interactions in the earlier years of my career with a host of international experts like Kellie O' Neill of USDA, Washington, J.D. Hood of Cornell, A. Bournier of France, K. Sakimura of Hawaii, Zur strassen of Germany, Hermann Priesner of Linz, Austria, J.C. Faure of South Africa, Jarsolav Pelikan of Czechoslavakia, C.B. Williams of England, to mention the major expertise on this group, In particular, I owe my deep sense of gratitude to Hermann Priesner without whose interaction my studies on the biodiversity of thrips would not have been possible. This article is therefore dedicated to the memory of these international experts.

References

Ananthakrishnan, T.N. 1956. Thysanopteran fauna of some grasses. *Zool. Anz.* 157 : 130.

Ananthakrishnan, T.N. 1960. Thysanoptera from the Nilgiris and Kodaikanal hills. *J. Bombay Nat. Hist. Soc.* 57 :

Ananthakrishnan, T.N. 1961. Allometry and speciation in *Ecacanthothrips* Bagnall. *Proc. Biol. Soc.* Wash; 74 : 275-

Ananthakrishnan, T.N. 1963. The Terebrantia of the Indo-ceylonese region. *Treabia,* 26 : 73-

Ananthakrishnan, T.N. 1964. A contribution to our knowledge of the Tubulifera (Thysanoptera) of India. *Opuscula Ent. Suppl.* 25 : 1-120.

Ananthakrishnan, T.N. 1967. Allometric growth patterns in some Tubulifera (Thysanoptera) *Indian J. Ent.* 29 : 81-

Ananthakrishnan, T.N. 1968. New gall thrips from India *Senckerberg Biol.* 50 : 179-194.

Ananthakrishnan, T.N. 1968. Patterns of structural diversity in males of some phlaeophilous Tubulifera. *Ann. Ent. Soc.* France 4 : 413-

Ananthakrishnan, T.N. 1969. Indian gall Thrips. *Marcellia* 36; 76-103 (with Jaglish).

Ananthakrishnan, T.N. 1971. Thrips in Agriculture, Horticulture and Forestry - Diagnostic, Bionomics and control. *J. Sci. Ind. Res. CSIR,* (30) 3 : 113-146.

Ananthakrishnan, T.N. 1973. *Thrips : Biology and control.* Macmillan, New Delih, 120 pp.

Ananthakrishnan, T.N. 1974. The distribution and host range of predatory thrips. *Indian J. Plant Protection,* 4 : 67-78.

Ananthakrishnan, T.N. 1976. Studies on *Mesothrips. Oriental Insects.* 10 : 185-214.

Ananthakrishnan, T.N. 1979. Biosystematics of Thysanoptera. *Ann. Rev. Ent.* 24: 231-242.

Ananthakrishnan, T.N. 1980. Thrips. In *Vectors of plant pathogens, Academic Press,* 149-164.

Ananthakrishnan, T.N. 1982. *Thrips and pollination biology. Current Science* 51: 168-172.

Ananthakrishnan, T.N. 1984. *Bioecology of Thrips.* Indira Publishing House. Michigan, 233 pp.

Ananthakrishnan, T.N. 1985. *Biology of gall Thrips.* In *Biology of gall insects,* Oxford IBH, New Delhi : 107-127.

Ananthakrishnan, T.N. 1986. *Bioresources Ecology.* Oxford and IBH, New Delhi, 172 pp.

Ananthakrishnan, T.N. 1988. Dynamics of the litter ecosystem in natural and interfered forest. *Int. J. Ecol. Envir. Sci.* 14 : 61-66.

Ananthakrishnan, T.N. 1996. *Forest litter insect communities : Biology and Chemical Ecology.* Science Publishers, Lebanon, USA.

Ananthakrishnan, T.N. 2005. Perspective and Dimension of phenotypic plasticity in insects. In *Insect Phenotypic Plasticity,* Science Publishers. Enfield, USA : 1-21.

Ananthakrishnan, T.N. and S. Sen 1980. *Taxonomy of Indian Thysanoptera.* Zool. Survey of India. Calcutta, Handbook Series : 1-234.

Ananthakrishnan, T.N., Mohan Daniel and N. Sureshkumar 1982. Seasonal and spatial distribution patterns of some phytophagous thrips infesting *Ricinus communis* (Euphorbiaceae) and *Achyranthes aspera* (Amaranthaceae). *Proc. Indian National Science Academy.* B48: 183-189.

Ananthakrishnan, T.N. & A. Raman 1989. *Thrips and gall dynamics* Leiden, *E.J. Brill,* Netherland, 120 pp.

L.A. Mound 2005. Fighting, flight and Fecundity : Behavioural determinants of Thysanoptera and structural diversity. *In Insect Phenotypic Plasticity,* Science Publishers, Enfield, USA : 81-105.

Lewis, T. 1973. *Thrips : Their Biology, Ecology and Economic importance.* Academic Press, London, 349 pp.

Mani, M.S. 1964. *The Ecology of plant galls.* Walter Junk Publisher, The Hague, 434 pp.

Mani, M.S. 1973. *Plant galls of India.* Macmillan India, New Delhi, 353 pp.

Ramakrishna, T.V. 1928. A contribution to our knowledge of the Thysanoptera of India. *Mem. Dept. Agric. India* (Ent. Series) 10; 215-316.

Raman, A. and Ananthakrishnan, T.N. 1985. Cecidogenous *Crotonothrips - Memecylon* interaction. *Proc. Indian Acad. Sci.,* 95 : 103-120.

Varadarasan, S. and Ananthakrishnan, T.N. 1981. Biological studies on some gall thrips. *Proc. Indian Nat. Sci. Acad.* B. 48 : 35-43.

Chapter - 11

Ecological Diversity and Insect Biodiversity in Cold Deserts in India

G. Thirumalai, Kailash Chandra and Avtar Kaur Sidhu*[@]
Southern Regional Centre, Zoological Survey of India, Chennai
email: thiruambi@gmail.com
*Central Zone Regional Centre, Zoological Survey of India, Jabalpur
@High Altitude Regional Centre, Zoological Survey of India, Solan

Arid and semi-arid region covers 38.8% of India's total geographical area and spread over 10 states and the cold arid zones are located in Trans-Himalayan region covering 5.62% of India. The Trans- Himalayan biogeographic zone is located across the divide of the Great Himalaya. In northern India this zone comprises almost all of Ladakh and Kargil districts (J&K), parts of Lahaul & Spiti and Kinnaur districts (Himachal Pradesh) and parts of Uttarakhand State. A large percentage of area in the Trans-Himalaya is high altitude cold desert (approx. 100,000 sq km) and is characterized by exceptionally low temperatures and precipitation.

Hot and cold deserts, no matter where they are located, are always dry. Deserts are all dry and arid places, but they are not always hot. Some deserts can be cold most of the year depending on their location. At night, hot air from the day dissipates and, in some areas, temperatures can drop below freezing. In hot deserts, the fragmenting force of temperature is slower. Rock surfaces reach 70°C or 80°C at midday and cool down to freezing point at midnight. Expansion under the sun and contraction at night weaken the surface layers and cause flaking. Heat and cold produce the least observable effect in sandy deserts. In contrast, their impact is much greater in rocky deserts. In cold desert, rainfall is frozen at night in winter. When water goes into the tiny cervices and expands, the rocks are forced to split up. Many hot deserts lie in the subtropical zones. Most cold deserts are found in higher latitudes and higher elevations. They are usually found between the subtropics and Polar Regions.

The cold deserts of India include Ladakh (J&K); Lahaul & Spiti, Kinnaur and Bharmour (H.P.); pockets of northern Uttarakhand and Sikkim. These regions have a unique physical, biological, hydrological and anthropological setting that is markedly different from that of the adjoining regions. Ladakh is also known as 'Mini Tibet'. Lying between 75°15' to 80°0' E latitude and 32°30' to 37° N longitude, Ladakh is spread over an area of 96,701 Sq Kms. This region remains closed during November to June every year as Srinagar-Leh and Leh -Manali highways connecting Ladakh with rest of the country get blocked due to heavy snow fall in the harsh winter conditions during that time.

Ladakh falls in the Trans-Himalayan region of the western Himalaya and is the northern most part of India. It is bounded on the north by Xinjiang, on the south by Lahaul and Spiti, on the west by Kashmir and on the east by Tibet. The borders of Ladakh touch countries like Afghanistan, Pakistan, China apart from Kashmir and Himachal Pradesh in India.

Climate

The climate of Ladakh is extremely dry and cold, approaching arctic conditions. Summers are short, though they are long enough to grow crops in the lower reaches of the Suru valley. In summer the weather is dry and pleasant, with average temperatures between 10-20°C, while in winter, the temperature may dip to -15°C, on an average. Daytime temperature in summer rises to 35°C but in winter, it remains below the freezing point. Minimum temperature during winter may be as low as -40°C in certain parts of Ladakh. The regions on the north flank of the Himalayas-Dras, the Suru valley and Zanskar-experience heavy snowfall and remain virtually cut off from the rest of the country for several months in the year.

The plant and animal species that manage to exist in this region are well adapted to the harsh, extreme environmental conditions. Compared to the main Himalaya, this region is floristically impoverished. Floristically, Ladakh comes under 'Alpine Desert Flora', and its natural vegetation can be described as comprising of shrubs, thorny bushes and herbaceous plants. Natural tree elements are largely lacking, except for more humid regions of Nubra valley and other river valleys. The most common trees in the area are the Poplar and Willows. Arid alpine pastures in Ladakh are dominated by low thorny scrubs, Tibetan furze and a variety of grasses.

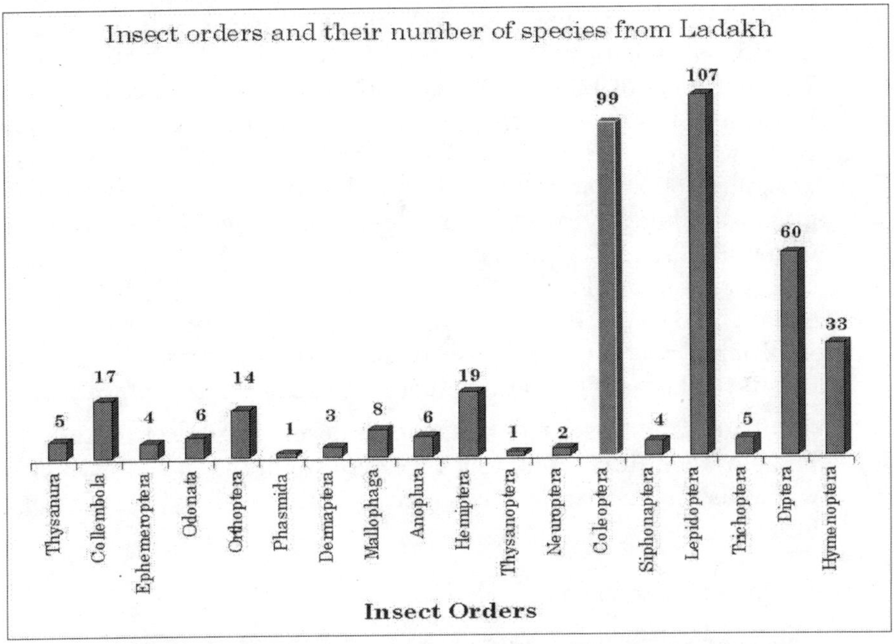

Floristically, this region has affinities with the main Himalaya and diverse regions such as Afghanistan, Tibet, the Mediterranean region and the Indian desert areas. About 750 species of flowering plants have been reported from the region. Unique floristic features of this region include a large number of endemics, a high proportion of monocots and the absence of large tree species. In this region, there are a number of species of great medicinal and economic importance and as a result have great conservation importance. Wild relatives of numerous species including onion (*Alium* spp.) rhubarb (*Rheum*) and barley (*Horedum)* are found in the Trans-Himalaya. It is believed that over 50% of plants in the region have recorded medicinal value.

Despite low plant diversity and productivity, the fauna is surprisingly diverse and has elements from the Himalayan as well as Mediterranean and Tibetan regions. Invertebrate diversity in the region is largely undocumented and still remains to be explored in detail. A number of surveys were conducted in the Trans -Himalayan regions during the seventies by Dr M.S. Mani, followed by the Wild Life Institute of India in 1994 and Zoological Survey of India, in 1999, 2005, and 2008.

Wetlands

High altitude wetlands can be broadly separated as (a) high altitude arid zone wetlands and (b) wetlands resulting from a monsoonal moisture cycle as seen in Kashmir and further east. The wetlands of Ladakh are typical of high altitude arid zones. Ladakh is bestowed with some very large brackish water lakes namely, Pangongtso, Tso Morari and Tsokar. The area also harbours some vast freshwater marshes in the river valleys. The great salt lakes fringed with marshes such as Hanle, Harong, Chushul, Tsokar, Tsomoriri provide, during the brief summer, conditions suitable for the breeding of one of India's most endangered species of avifauna, the black-necked crane. Additional species which use this habitat for breeding are the bar-headed geese, Brahminy ducks and some species of waders.

The faunal resources of this cold desert exhibit Oriental and Palaearctic affinity. The rich invertebrate fauna of the marshes, wetlands and agroecosystems attract a wide variety of migratory birds.

Entomofaunal Diversity

The cold desert insect fauna called nival insects is an ecologically specialized cold adapted mountain dwelling group occurring beyond the timber line above an elevation of 3000 meters and up to 6000 meters above sea level (Mani and Singh, 1961). Nearly 392 nival insect species belonging to 18 orders hitherto reported from Cold Desert of Ladakh (Table 1).

Table 1. List of insect orders and the number of species from Cold Desert of Ladakh.

S. No.	Order	Number of species	Percentage
1	Thysanura	3	0.76 %
2	Collembola	17	4.3 %
3	Ephemeroptera	4	1.0 %
4	Odonata	6	1.5 %
5	Orthoptera	14	3.5 %
6	Phasmida	1	0.25 %
7	Dermaptera	3	0.76 %
8	Mallophaga	8	2.0 %
9	Anoplura	6	1.5 %

10	Hemiptera	19	4.8 %
11	Thysanoptera	1	0.25 %
12	Neuroptera	2	0.50 %
13	Coleoptera	99	25.5 %
14	Siphonaptera	4	1.0 %
15	Lepidoptera	107	27.2 %
16	Trichoptera	5	1.2 %
17	Diptera	60	15.3 %
18	Hymenoptera	33	8.4 %
	Total	392	

The order Thysanura is represented by a species of *Machilanus* and two species of *Ctenolepisma* in Ladakh region. All these species are known to localize in the Indus drainage area. Out of 250 species under 96 genera and 12 families of collembolans known from India, 17 species belonging 10 genera under seven families are reported from Ladakh region constituting more than 50% species are localized to Trans-Himalayan region. So far, four species belonging to two families of Ephemeroptera are known to occur in Ladakh region. Knowledge of the order Odonata occurring in the cold desert is at present very scanty. Presently six species belonging to four genera under family Libellulidae are recorded from Ladakh. All these species are common in the valleys of the Ladakh region. Order Orthoptera includes 14 species of the families namely Acrididae and Tettigonidae from Ladakh. More than 50% species are endemic and represented mainly by the Palaearctic elements. There are also Mediterranean forms belonging to the genera *Sphingonotus and Oedipoda.* The central Asian elements are represented by one species of *Gomphomastax,* a genus which is known to be derived from ancient tropical and subtropical subfamily Eumastacinae. Though the great bulk of the species of Orthoptera from the NW. Himalaya seem to be generally localized at elevations of above 3500 m., the maximum altitudinal record of 5000 m. above msl is reached by the Tettigonid, *Hyphinomus fasciata* Uvarov. The Leaf insects and Stick-insects is so far represented by recently described stick insect *Ladakhomorpha longipes* Hennemann & Conle, 1999, belonging to family Pachymorphidae. The order Dermaptera is represented by three species of the Palaearctic genus *Anechura,* belonging to the mountain-inhabiting subfamily Anechurinae of family Forficulidae. Alfred *et al.* (2003) mentioned the record of eight species of Mallophaga belonging to five genera of Mallophaga from Ladakh. Recently, *Syrrhaptoecus tibetanus* Waterston of the family Philopteridae has been recorded on feathers of Tibetan Sandgrouse (*Syrrhaptes tibetanus* Gould) from Ladakh region. The Order Anoplura of the Ladakh region is represented by eight species belonging to three families. The Order Hemiptera constitutes about 5% of the total insect fauna of cold desert region and comprises mostly Lygaeidae including, two endemic genera *Dalmacoris* Hutchinson and *Tibetocoris* Hutchinson of the family. Altogether, 19 species under 9 families are represented from Ladakh. Presently only one species of Thysanoptera, *Thrips tabaci* Lindeman of the family Thripidae has been found attacking different vegetable crops in cold desert region of Ladakh. The neuropterans are represented by two species, *Myrmecaelurus punctulatus* collected at an altitude of about 3450 m. from Leh in the Indus drainage area, (also known from S. Russia and Hungary) and *Chrysopa carnea* as natural enemy of fruit pest in Ladakh. The order Coleoptera represents about one third of cold desert insect fauna of the Ladakh region. Hitherto, 99 species of beetles belonging to 16 families are known from Ladakh. Most of the species are Palaearctic with some elements of Mediterranean and Nearctic region. Four species of Fleas belonging to two genera of the family Ceratophyllidae are so far recorded from Ladakh. All these species are collected from bodies of small mammals that are widely distributed in the Indian subcontinent. Sixty three species of butterflies belonging to five families are being reported from Ladakh and most of the species are Palaearctic. Family Papilionidae is represented by two genera *Papilio* and *Parna-ssius*. The Holarctic species *Papilio machaon* known from Ladakh is also distributed in Tibet. The Holarctic genus *Parnassius* is represented by 11 species and about 18 subspecies, of which more than 80% are endemic. Out of 63 species of butterflies so far known from Ladakh, presently ten species/ subspecies are included in the list of threatened categories of Indian Wildlife (Protection) Act, 1972 and IUCN. However, as it is known that due to very scarce vegetation in the Cold desert of Ladakh, the population of butterflies is very less and many species/ subspecies are endemic to the region or localized to north-west Himalayas. Despite the rarity of these species, the specimens are collected illegally for the international trade due to high value. The butterflies collected from Ladakh, Lahaul & Spiti, Sikkim and Meghalaya fetch about 100 million US dollars in international trade. Forty four species of moths representing 12 families are known from Ladakh. The family Noctuidae includes 14 species, followed by the Arctiidae (6) and Pterophoridae (5). Yponomeutidae is represented by one species, which is recorded as a pest of plantations. The family Pterophoridae includes 5 species from Ladakh. Family Sphingidae commonly known as hawk moths also possess diurnal species. Two species namely *Hyles nervosa* Rothschild and Jordan and *Hyles laggi* Rothchild are known from Ladakh and the later species is the first report from Ladakh. Both the species occur at an elevation of 4500 m. There are several species of moths, which are also recorded as pest of forests, vegetable crops and plantations in cold arid region of Ladakh. It was reported that *Helicoverpa armigera* Hubner, *Pieris brassicae* Linnaeus, *Plutella xylostella* Linnaeus, *Phyllotreta* spp. *Agrotis ipsilon* Hubner and *Etiella zinkenella* Treitschke were found damaging the different vegetable crops, while gypsy moth (*Lymantria* spp.), goat moth (*Cossus cossus* L) and ermine moth (*Yponomeuta rorellus* Hubner) were found to be causing damage to poplar and willow plantations. Five Palaearctic species of Caddis flies belonging to two families Sericostomatidae and Limnophilidae are so far known from Ladakh. The genus *Pseudostenophylax* (=*Stenophylina*) is endemic to the region and localized in the Indus valley. All the species are reported between 3000 m and 4640 m altitude.

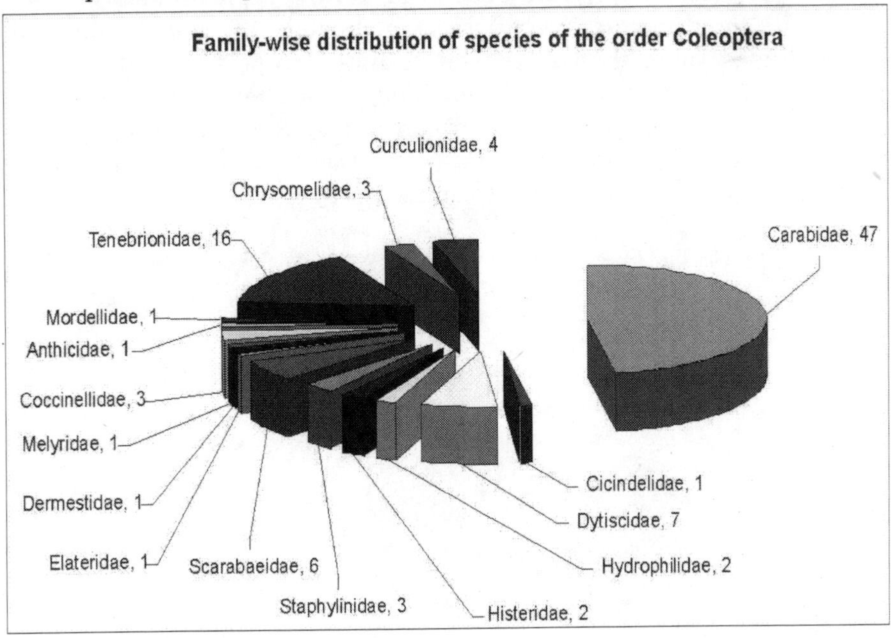

Family-wise distribution of species of the order Diptera from Cold Desert of Ladakh

S. No.	Order/ Suborder/ Family	No. of Species
	Order Diptera	
	Suborder Nematocera	
1	Family Culicidae	2
2	Family Chironomidae	51
	Suborder Brachycera	
2	Family Asilidae	1
3	Family Syrphidae	1
4	Family Tephritidae	2
5	Family Ephydridae	3
		60

The present state of our knowledge on Diptera from Ladakh region is very poor. During the survey in 2008 by Zoological Survey of India, several specimens of flies are collected up to 5000 m above from sea level. The Chironomidae of Tso Morari Lake (Ladakh) was studied and 51 species were reported. A species of the family Syrphidae *Episyrphus balteatus* has been reported as a natural enemy of important fruit pests in arid cold region of Ladakh. Two species of family Tephritidae were also collected from different localities during 2008 survey by ZSI. Besides, more than 20 species of dipterans belonging to families Tipulidae, Simuliidae, Bibionidae, Stratiomyiidae, Tabanidae, Phoridae, Calliphoridae, Anthomyiidae, Sarcophagidae, Muscidae and Tachnidae were also collected from Ladakh region, which are in process of determination by ZSI.

Family-wise distribution of species of the order Diptera

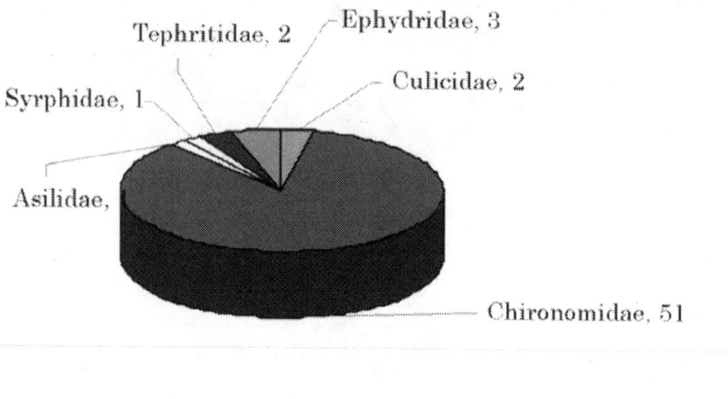

Cold desert area in Razang La

Pangong Tso, a brackish water lake, 5200m above msl

Cold desert area in Razang

Parnassius epaphus hillensis from near the Khardung La at 5100 m msl.

An unidentified pentatomid

The cowpea aphid, *Aphis craccivora* Koch

Sphingonotus sp

An unidentified Ephemeroptera

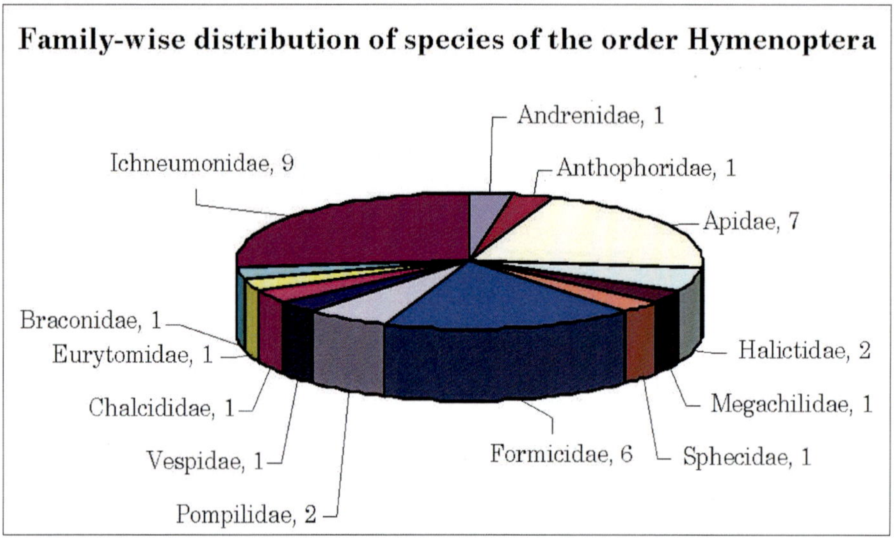

Hymenoptera, with 33 species so far recorded, constitute about 10% of the total nival insects. Of the thirteen families known from Ladakh region, the families Apidae, Ichneumonidae and Chalcidae are the most abundant. The species of Formicidae were recorded up to 5000 m asl. In cold desert of Ladakh, almost every locality of study area, showed the presence of family Apidae, paticularly the genus *Bombus*. The genus *Apanteles* sp of the family Braconidae has been recorded as natural enemy of agriculture pests. The family Ichneumonidae is represented by 9 species belonging to major genera viz. *Acrotemes*, *Prothemes*, *Ophion*, *Ctenichneumon*, *Ichneumon*, *Netelia*, *Lissonota*, *Meringopus*. The Superfamily Chalcidoidea includes several families, which parasitize the dipterans and also visit the flowers during the adult stages. During the recent rapid assessment survey of Ladakh by the Zoological Survey of India, more than 15 species belonging to Superfamily Chalcidoidea were collected. However, on their inclusion the number of known species goes to 48.

Family-wise distribution of species of the order Hymenoptera from Cold Desert of Ladakh

S. No.	Order/ Suborder/Family	No. of species
	Order Hymenoptera	
	Suborder Apocrita	
	Infraorder Aculeata	
1	Family Andrenidae	1
2	Family Anthophoridae	1

3	Family Apidae	7
4	Family Halictidae	2
5	Family Megachilidae	1
6	Family Sphecidae	1
7	Family Formicidae	6
8	Family Pompilidae	2
9	Family Vespidae	1
10	Family Chalcididae	1
	Infraorder Parasitica	
11	Family Eurytomidae	1
12	Family Braconidae	1
13	Family Ichneumonidae	9
	Total	33

Discussion

The exclusively mountain autochthonous faunal elements which arose *paripassu* with the Himalayan uplift are mostly composed of Palaearctic elements, the forests in particular are of Turkmen subregional affiliation in the high altitude cold deserts of the Himalayas. The cold adapted fauna lacks tropical Indian, south Chinese, Indo-Chinese and Malayan derivatives. The bench mark of Ladakh fauna is of the Tibetan plateau which has the admixture of Oriental and Palaearctic elements. Animals in this cold desert area are sparsely distributed under harsh environmental conditions on rocky terrains where they find poor shelters and scanty vegetation. The major chunk of Invertebrates is mostly endogeous, terricolous with an increasing degree of endemism from west to east. Permanent existence of Collembola, Diptera, Acarina and Salticid spiders at the highest altitude of 6800m has been reported by Mani (1962, 1968) and Swan (1961). Quite interesting is the observation of Mani (1973) that *Agraptocoris margaretae* (Hutchinson) occurs at an elevation of 5400m, the highest altitude record for the order Heteroptera (family: Miridae) from the world. The earlier known "devoid of life" arctic zones of the Himalaya do have a home for a unique community of organisms. The present day endemic faunal constituents possibly are the descendents of Middle Asian ancestral stocks.

References

Alfred J.R.B., Tak, P.C. and Sharma, R.M. 2003. Faunal Resources of Ladakh, *ENVIS Newsletter*, 10 (1-2): 12-16.

Chandra, K and Sindhu, A.K. 2009. Insects of Ladakh. *ENVIS Newsletter*. **15**(1&2): 5-10.

Fotedar, D.N. and Bhagat, R.C. 1984. An annotated Bibliography (1848 to 1983) Jammu Kashmir and Ladakh (Zoological Studies), pp 128. International Books and Periodical Supply Service, New Delhi.

Maheswari, G and Maheswari, G. 2004. Insect Biodiversity of Ultraoligotrophic Lakes of Ladakh Himalaya, India, *Advancement in Insect Biodiversity*. (Editor, R.K.Gupta), *Agrobios* (India), Jodhpur: 109-122.

Mani, M.S. and Singh, S. 1961. Entomological Survey of Himalaya, Part XXVI. A Contribution to our Knowledge of the Geography of the High Altitude Insects of the Nival Zones from the North-West, Himalaya. *J. Bombay nat. Hist. Soc.*, **58**(2): 387-406.

Mani, M.S. 1962. Introduction to High Altitude Entomology. London: Methuen and Co. pp. 306.

Mani, M.S. 1968. *Ecology and Biogeography of High Altitude Insects*. The Hague: Dr. W. Junk Publishers. pp. 527.

Mani, M.S. 1973. Biogeography of the Himalaya. *In*: *Ecology and Biogeography in India*. Edited by M.S. Mani. The Hague: Dr. W. Junk Publishers. pp 664-681.

Swan 1961. The ecology of the High Himalaya. *Sci. Amer*. **205**(4): 68-78.

Chapter - 12

Insect functional dynamics in the unique xeric ecosystem of The Thar Desert, northwestern Rajasthan (India): Before and after the advent of Indira Gandhi Nahar Pariyojana

Dr. B.K. Tyagi

Centre for Research in Medical Entomology (ICMR), 4-Sarojini Street, Chinna Chokkikulam, Madurai 625002, TN, India Email: bk_tyagi@sify.com

INTRODUCTION

The Great Indian Thar Desert, merely 10,000 years old, is one of the youngest deserts in the world, implying that it had been subjected to continuous transformation under natural or man-made physiographic stresses (Warren *et al.*, 1996). During yore, the Thar Desert region had been the abode for the once existing historic Harappa civilization in Kalibanga along the extinct holi river Saraswati, traversing across the Great Indian Thar Desert, together with Yamuna and Ghaggar rivers. At present only the relic Ghaggar marks its presence during monsoon, while Yamuna has in course of time changed its direction, now flowing far away from Rajasthan's north-eastern border into the neighbouring State of Uttar Pradesh. *Inter alia* several different kinds of changes took place in the Thar Desert region, characteristically experiencing a more or less set frequency of decadal droughts leading to emigration of local folks across the border to other States with high disease endemicities (*e.g.*, Punjab, Gujarat, Madhya Pradesh and Uttar Pradesh etc.), intermittent floods in the lower reaches of the Luni river basin particularly Barmer district and adjoining districts in the Jodhpur zone, and the excessive recurrence of the sand-storms resulting into expansion of the desert. Yet, the greatest discernible change the Thar Desert seemed to have undergone with a human interruption with nature is the provision of extensive canalization through Indira Gandhi Nahar Pariyojana (IGNP), aided with other two canal systems, viz., the Bhakra Sirhind feeder and the Gang Canal, which is probably world's one of the largest canal systems of its type in a desert ecosystem. Though, originally developed to boost the future prospects of agriculture in the virgin deserts and animal keeping through planned and sustained irrigation, nevertheless, due to multiple reasons associated with canal-water management, the insect fauna of the region also exhibited an enormous shift in its ecosystem-related characteristics (Tyagi & Chaudhary, 1997). The Thar Desert insect fauna was minimal and monotonous during the time before its apocalyptic metamorphosis in the mid-20th century. While existence of several insect species is at risk due to destruction and/or catastrophically modified habitats, there are other species who have not only ushered into the Thar Desert with the advent of the IGNP but are also thriving on the cost of other local species by way of eroding their domains competing for space, food, breeding and survival. Consequently, the Thar Desert has turned out to be a living model to demonstrate as to how apocalyptic vicissitudes in nature and environments can bring about subtle transformation into both the faunal composition and ecosystem-related cadence.

DESERT PHYSIOGRAPHY AND TOPOGRAPHY: BEFORE AND AFTER THE ADVENT OF THE THREE MAJOR CANAL SYSTEMS (INCLUDING IGNP) IN THE THAR DESERT

The geographic limits of the Thar region are generally understood to be between the irrigated lands of the Indus and Sutlej and the western edge of the Aravalli Hills (i.e., approximately between 24°-30° N and 68°-78° E), although in Rajasthan State where the desert in the country is maximally present, it is spread over nearly 75,000 km^2 (62% of the total land of the State) and harbouring a population of 12.8 million people (39% of the population of the State). The seasonal Luni river divides the desert into two parts; in the north a vast stretch of sand dunes or levees and in the south an expansive basin rimmed by flat- topped hills of sandstone and limestone. The climate of the Thar Desert is, generally speaking, of the tropical desert type with arid hot environment. May is the hottest month with mean maximum temperature at 41°C in the plains of the desert, although in past it sometimes soared to as high as 50°C. In contrast, January is the coldest month in the Thar desert., but it had seldom been recorded as low as –5.9°C. Monsoon had been highly deficient and erratic with humidity ephemeral; the average rainfall being around 200 mm. The relative humidity is generally very low during the hot and dry periods of the year; during summer it dips to as low as 15-34%, making the summer very dry and hot.

Some of the most significant events in the Thar desert had been the frequent occurrence of dust-storms, famines and draughts; the last one being most during 1901 and 1939. Population was thin (6 per sq. km) in the Thar Desert.

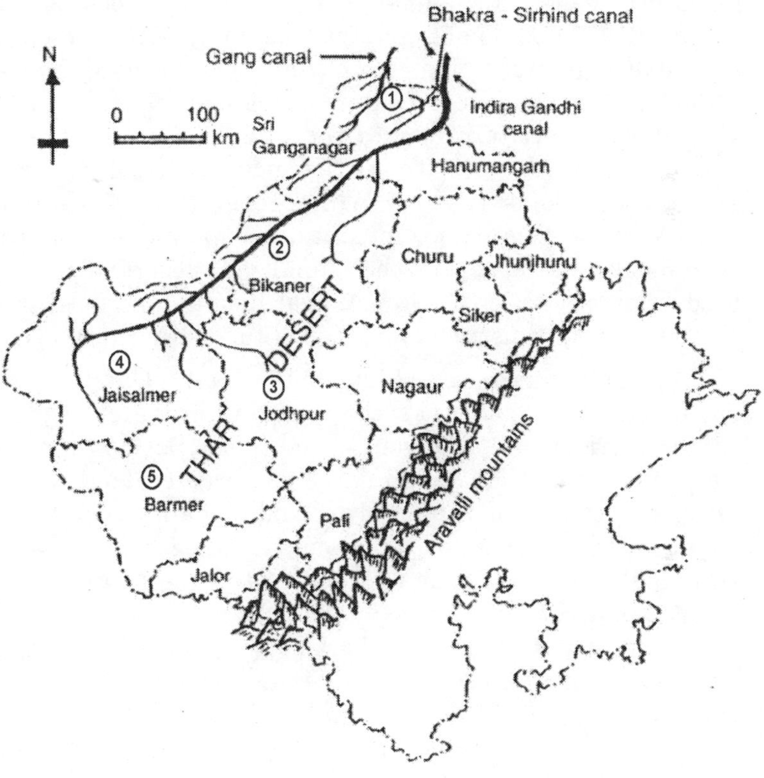

Fig. 1. Twelve desert districts in the Thar Desert. Those districts depicted by encircled numerals have experienced repeated *P. falciparum* dominated malaria epidemics; all falling under canal based irrigated, particularly, IGNP.

Three major canal systems exist in the Thar Desert region today. The first is the Gang canal project which became fully operative in 1928, with its 1251 km distribution system in Sri Ganganagar district only. Along with the Bikaner canal feeder, it provides water to about 65% of the 300,000 ha of cultivable land. The Bhakra-Sirhind canal project, with its 1219 km long distribution system, was started in 1955 to utilize a further

300,000 ha to cultivable land. The most important of all the canal systems is the Indira Gandhi Nahar Pariyojana project (IGNP) which is one of the world's largest projects of its kind in the desert. It started in 1958 and irrigation of part of the area began in 1961. The project, when completed, will include a 649-km long main canal (of which 445 km is in the Thar Desert) and over 9000 km of distributaries. Although only two of the three stages envisaged are completed by now, the main canal including feeder and 5500 km of distributaries is already irrigating 950,000 ha of land (approximately 50% of the irrigated area that is planned. This Herculean task, bringing Himalayan waters to the Thar Desert, aims at transforming vast stretches of western desert terrain into verdure. As a result, the annual agriculture production from the newly irrigated land now exceeds Rs. 6500 million. While several irrigation-intensive crops such as sugarcane, mustard, groundnut, etc., have been established in some of the canalized desert districts, the production of rice in Sri Ganganagar district, the most irrigated among all desert districts, has topped the list in the whole State.

At many places along the main Indira Gandhi canal, in particular, the problem of seepage water and water-logging have created serious mosquitogenic conditions. Besides the surplus irrigation water that accumulates on irrigated land, vast areas along the course of the Indira Gandhi main canal have been engulfed by seepage water from the irrigation canals on a perennial basis. An estimated 8600 ha in the Nalibed area, Sri Ganganagar/Hanumangarh district, is permanently inundated under the IGNP Stage I by the Ghaggar river floodwater while 1000 ha is converted into marshy land due to water- logging and excessive seepage from the canal in Stage II. Supply of canal water to the desert region over several decades has caused a rise of the water table and water-holding potential of the soil, and a plentiful growth of vegetation, particularly hydrophytic weeds. Such extensive breeding grounds have proved to be favourite sites for breeding by several different vector mosquitoes, but also of dragonflies, many of which were even not found in the interior of the Thar Desert.

RAINFALL, AIR TEMPERATURE AND RELATIVE HUMIDITY CHANGES IN THE THAR DESERT'S IGNP COMMAND AREA

A recent compilation of percentage departure of monsoon season rainfall (June to September) for the period 1901–2000 over west

Rajasthan by Dr. S.M. Kulshrestha (*pers. comm.*) depicts how significantly the drift has been over past 100 years (Table 1). Ramakrishna & Rao (1991) have found an increase in the mean decadal rainfall in the district, which had been subjected to canal-based irrigation for a considerable period. Even more recently, Sikka & Kulshrestha (2001), have attempted to draw relationship between climate and health, including vector-borne diseases, especially malaria.

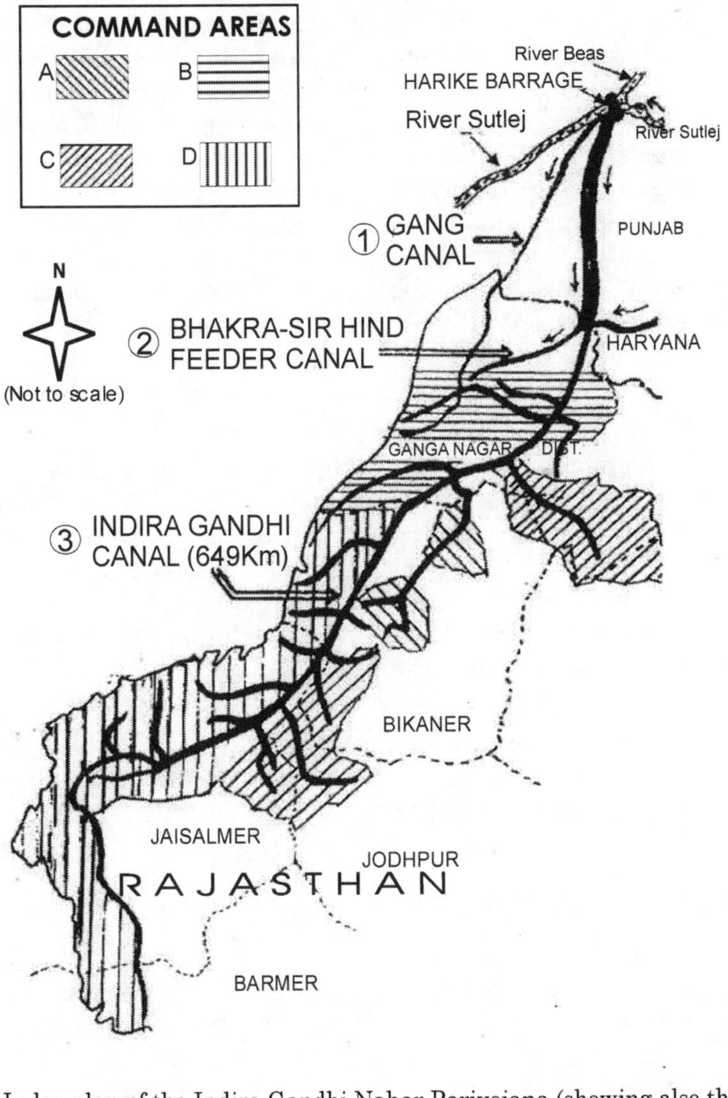

Fig. 2. Index plan of the Indira Gandhi Nahar Pariyojana (showing also the Gang canal and the Bhakra-Sir Hind Feeder canal).

The Gang canal and IG canal projects aimed to utilize 7.59 MAF of water from Ravi-Beas duo waters to irrigate an 1.143 mill-ha area. In the Ganganagar district, the Bikaner district and the Jaisalmer district, 708, 775 ha, 37,022 ha and 84 ha of land, respectively, has been put under irrigation (Roy, 1983). The impact of irrigation in Rajasthan is seen from the vegetational and micro-climatic changes that have occurred in the region. Sri Ganganagar receives an annual rainfall of 243 mm. The annual rainfall of Sri Ganganagar showed an increasing trend, at the rate of 1029 mm/year, during the period 1926-1993. The mean annual rainfall of Bikaner is 286 mm with a coefficient of variation of 48%. The region experienced droughts once every 3 years (38% chance). The rainfall (annual) at Bikaner showed an overall decreasing trend at a rate of 0.147 mm/year during 1926-1993, which is forcibly attributable to consecutive droughts which occurred after 1961. Jaisalmer receives a mean annual rainfall of 185 mm with a coefficient of variation of 64%. The rainfall has until recently been erratic and poorly distributed with droughts occurring once every 4 years (24% chance). However, the overall trend in annual rainfall at Jaisalmer which remained the same during 1901-1993 seems to be shifting towards an escalation in recent years.

The ambient temperature, too, has shown decreasing trend being 0.039°C/year at Sri Ganganagar, 0.023°C at Bikaner, and 0.009°C/year at Jaisalmer. With regard to relative humidity in the Thar desert, Rao (1996) recorded 5-10% higher humidity measurements in cropped surfaces at 1-m height in the irrigated villages compared to the unirrigated cropped surfaces. This highlights the fact that the amount of canal water available in the desert region can influence atmospheric humidity.

These observations sufficiently indicate towards a shift in the abiotic characteristics of the Thar Desert under the IGNP command area as compared to that in the typical desertic environs.

INSECT FAUNA IN THE PRE-IGNP ERA IN THE THAR DESERT

As is expected from a inhospitable and hostile desertic environment, but also the monotonous insect faunal life, of the Thar Desert, in fact very few scientific explorations were made in this unique ecosystem. Most of these researches revolved around the migratory locusts which used to enter the Thar

Desert from Pakistan and cause extensive damage to the crops in the Indian part of the Thar Desert in northwestern Rajasthan. Such was the havoc created by this locust species to agriculture that a specific Locust Research Centre had to be stationed at Jodhpur for developing ways and means to cease the orientation of this menace and its control by forecasting to farmers. Another insect which has inhabited the Thar Desert for centuries is the dung beetle (genera *Scarabaeus* and *Gymnopl-erus*) which play an important role in pasture ecosystem where they annually breakdown tones of animal dung and much of it incorporated into the soil to increase fertility and also to clean the environment. Pruthi & Bhatia (1952) lamented dismal insect faunal surveys in the Thar Desert particularly interior rural environs. Even whatever mosquitoes and sandflies reported from the desert areas were actually originating from the urban settlements (Jaswant Singh, 1933). This vacuum of knowledge largely attributed to the monotonous life forms of insects in the Thar Desert where they were largely innocuous, although ectoparasites on camels and sheeps could have been aplenty and must be causing health concerns to these desert milching animals.

THE IGNP COMMAND AREA CHARACTERISTICS AND VARIOUS EVENTS DIRECTLY OR INDIRECTLY ASSOCIATED WITH FACTORS ACCENTUATING INSECT DIVERSITY IN THE THAR DESERT

The Indira Gandhi Nahar Pariyojana, conceived in 1948, began to be constructed in 1958, and the first waters were flown in 1962. Extensive canalization with improper canal linings caused at many places huge areas flooded with stagnant waters. The IGNP has substantially changed the ground water hydrology in the command area. Before the project, in 1952, the ground water table in stage-1 was observed between 40-51 mbgl. Compared to this, in 1973, the water table was observed between 9.00 and 32.00 mbgl, which considering soil types available in the Thar Desert, resulted to large scale waterlogging due to rise in water table. The perennial presence of water sheet and the resultant greenery decreased ambient temperature (-0.52°C for 100 years) (Pant & Hingane, 1988), and increased rainfall (Ramakrishna & Rao, 1991). Ironically duststorms, once so frequent, became occasional as even events of droughts and famine decreased. Years of sustained canal water supply to the Thar Desert region changed the crop pattern as well. Till the turn of the first half of the 20th century Sorghum or Jwar (*Sorghum vulgare*), bajra or pearl millet (*Pennisetum* typhoides) moth (*P. aconifolius*), moong (*P. radiatus*) and gwar (*Cyamopsis psaraliodes*) etc. were the main desert crops completely dependent on the monsoon-fed rains. However, with the advent of extensive canal irrigation in the sixties, some irrigation crops got ushered into the Thar desert, such as cotton, wheat, mustard, groundnut, sugarcane and, in particular, paddy have also thrived making Sri Ganganagar in the Thar Desert the number one district in the Rajasthan State for rice production. With the groundwater quality changing, the soil, properties which earlier was coarse-textured with low fertility, have also altered making the land more irrigable and productive.

Obviously the Thar Desert physiography was getting transformed to a new ecosystem within the vast desert region which was to harbour a number of insect species for which the Thar was a *terra incognito* so far.

INSECT FAUNA OF THE POST-IGNP PERIOD

A large number of insect orders are represented in the Thar Desert today, prominent amongst which are: dragonflies, aphids, orthopterans, termites, butterflies, coleopterans, flies including mosquitoes, sandflies and tabanids. Some of these prominent insects are described here.

(1) Termites: In spite of the fact that 2/3rd of the State is dry and arid, yet the termite fauna in arid ans semi-arid regions of Rajasthan is quite rich. Out of nearly 2865 species worldwide, 348 occur in India and 34 in the Thar desert (Rathore & Mandal, 2005). Thar Desert abound in termite species now with the availability of moisture in soil. *Psammatermes rajasthanicus* prefers sandy areas with little vegetation, while *Anacanthtermes sakerensis*, the harvester termite, is found in sandy and gravelly soil both near cultivated and uncultivated areas. A number of species have exhibited a great deal of potential to adapt to new soil types following extensive canalization.

The biotic pressure in desertic Thar region is immense due to anthropogenic activities arising out of demographic explosion, also of animals. This situation obliterated the dismal vegetation of the desert and termites appeared subtly the *numero uno* pests of both the natural and man-made products containing mostly cellulose. To this the IGNP had contributed in its own way by providing more moisture to subterranean life, i.e., termites.

While, on one hand, desert-plant specific termite species are being threatened, others more dynamic species are adapting fast to expand their distribution in the Thar. The scenario can lead to the replacement of many a desertic termite species (e.g., *Incisitermes didwanaensis*) by those of mesic ones, such as *Amitermes belli*. Ironically, western Rajasthan in yore, i.e., about Pleiostocene period, sustained a richer, more varied and tropical termite fauna than today. Species like *Psammotermes rajasthanicus*, *Angulitermes jodhpurensis* and *Microcerotermes laxmi* certainly thrived in olden days; the last two being endemic to the Thar desert.

Fig 3. Damselfly, *Rhodischnura nursei* (above), and a gomphid dragonfly (below) perching and basking in the sun.

(2) Dragonflies: Dragonflies and damselflies (Order Odonata) thrive in the Thar Desert, particularly in areas where stagnant water bodies rife with aquatic plants exist. Most species encountered in the Thar Desert are the common ones, although one may occasionally come across an *Anax* sp, or a gomphid. The author has reported several species from around Kayalana lake in Jodhpur, with a quaint observation on *Rhodischnura nursei* Morton. This species likes to follow movements of a human being, tracking these precisely. Deeper in the desert where some monsoon-fed pockets occur during monsoon period, though rare, *Pantala flavescens* can be observed.

(3) Mosquitoes: The primitive Thar Desert was scantily having perennial water resources, therefore culicine mosquitoes were hardly ever present in the desert. However, since the potable water in variety was resourced in the Thar desert, at least one species, *Anopheles stephensi,* believed to have immigrated from the Middle-East and inherently a desert species, did make its mark in the underground "*tanka*" and underpond "*beri*" – the two domestic potable water reservoirs present in each village community (Tyagi & Yadav, 1996). While *beri* had a longer history in the Thar Desert, *Tanka* seemed to have existed only after mid-16th century. Needless to re-emphasize that *An. stephensi* is a serious vector of malaria. In surveys made in mid-1980s culicines like *Culex quinquefasciatus, Cx. pseudovishnui* and *Aedes aegypti* were often reported from the Thar desert. Later on the present author also recorded Cx. tritaeniorhynchus from the IGNP command areas.

Until about two decades from now the Thar Desert was hardly considered an important habitat for the malarial disease or its vectors. This was due essentially to very low prevalence of malaria and without any epidemic in the region. In the wake of such knowledge about malaria, no significant anopheline survey was ever made in the interior parts of the Thar Desert, and the fewer studies then directed on the subject have been from around the townships. Also, medical entomologists got more attracted to the forested and mountainous districts of the Udaipur zone in southern Rajasthan, being highly malarious. It is not surprising, therefore, that in the early days nearly all the major faunal surveys in the State were organized in the Udaipur zone districts only, and very few could have been made in the Thar Desert. Although, the first ever account of Indian anopheline was presented on a 'wall map' in 1912 by the Central

Malaria Bureau, at Kasauli (subsequently transformed first into malaria Institute of India and, finally, the National Institute of Communicable Diseases, Delhi), it became possible only in the late twenties when for the first time four anopheline species from the then Rajaputana State were put on record (Covel, 1927). Later in his *magnum opus* on 'Family Culicidae: Tribe Anophelini' in the classic 'Fauna of British India' series, now largely out of date, Christophers (1933) described in detail the taxonomy and bionomics of these four species, *viz.*, *An. stephensi*, *An. culicifacies*, *An. annularis* and *An. subpictus*, which were stated to be occurring in at least a few desert districts as well. The total number of taxa in the State was first raised to seven and subsequently to 16 (Puri, 1955), without any change in the strength of the desert fauna. According to Puri (1960), most of the 16 anopheline species were restricted to the two southern forested districts, Sirohi and Udaipur, contributing 56% of the total fauna from the State, while the Thar Desert had a meagre share of seven (44%) species from five of the then eleven desert districts. Wattal & Kalra (1962) offered a comprehensive taxonomic key, also State-wise, for the adults of anopheline mosquitoes known to be occurring in India till then. Interestingly, though Rao (1984) extensively dealt with the Indian anophelines, but no mention was made to either the Rajasthan State or the Thar Desert while discussing the regional distribution of the taxa. Recently Nagpal & Sharma (1995) have inventorized India's anophelines.

Two more species, *An. nigerrimus* (new to Sri Ganganagar) and *An. splendidus* (new to Rajasthan State), collected out of a total of seven species in Sri Ganganagar district, were added to the list of anopheles from the Thar Desert raising the total to nine (52.9%), with only two additions to the faunal strength of the state, including *An. d'thali,* new to Rajasthan, and elevating the fauna from the Thar to 10 (55.6%), and in the State to 18 (34.6% of the total anophelines recorded from India). Subsequently, *An. barbirostris* from Bikaner and *An. fluviatilis* from Jaisalmer have also been recoded making the latest inventory shooting to 12 (66.6% of the total taxa enlisted from Rajasthan. It is comprehensible from the distribution of taxa in each district, as given in Table 1, that there stands a good possibility of discovering more new records from the Thar Desert, particularly the unexplored and/or less explored north-eastern and southern desert districts, provided long term surveys are undertaken, though at the moment the Thar Desert remains under represented from districts' point of view.

Table 1. District-wise distribution of anophelines

District	Species collected
Sri Ganganagar	*An. culicifacies, An. stephensi, An. subpictus, An. vagus, An. annularis, An. nigerrimus, An. splendidus*
Hanumangarh	*An. culicifacies, An. stephensi, An. subpictus, An. vagus, An. annularis, An. nigerrimus, An. splendidus*
Bikaner	*An. culicifacies, An. stephensi, An. subpictus, An. vagus, An. annularis, An. nigerrimus, An. splendidus, An. barbirostris*
Jhunjunu	Not surveyed
Churu	Not surveyed
Sikar	Not surveyed
Jaisalmer	*An. culicifacies, An. stephensi, An. subpictus, An. vagus, An. annularis, An. nigerrimus, An. splendidus, An. d'thali*
Barmer	*An. culicifacies, An. stephensi*
Nagaur	Not surveyed
Jodhpur	*An. culicifacies, An. stephensi, An. subpictus, An. annularis*
Pali	Not surveyed
Jalore	Not surveyed

As far as the Thar Desert is concerned, sibling species have been confirmed to be occurring in *An. culicifacies, An. subpictus* and *An. fluviatalis* (Table 2). *Anopheles stephensi* does not possess any species complex. *Anopheles culicifacies* complex comprising 5 species (A,B,C,D and E) is the most important group since the malaria flare up in the Thar Desert seem to be influenced by the irrigation of the desert lands and it is confirmed that species A, which is a vector, occurs prodigiously in the canal-irrigated areas (Subbarao *et al.*, 1999). From the point of view of distribution of sibling species in different vector species complexes, the Rajasthan State is stratified in Division I and II. The northeastern districts of the Thar Desert like Sri Ganganagar, Hanumangarh, Churu, Jhununu and a part of Bikaner and Sikar are classified into Division I which is harbouring species A. The remaining districts in the Thar Desert are grouped inti Division II where species B and C are most common. Therefore, it is suggested that areas with species A and C need be thoroughly monitored for various behavioural characteristics to help design the control strategies more purposefully.

Table 2. Malaria vectors with species complexes in the Thar Desert

Species	No. of sibling species	Nomenclature assigned to sibling species	Sibling species found in India	Sibling species found in the Thar Desert
An. culicifacies	5	A, B, C, D and E	all	A,B and C
An. fluviatialis	3	S, T and U	all	Not known yet
An. subpictus	4	A, B, C and D	all	Not known yet
An. annularis	2	A and B	all	Not known yet

As regards *An. annularis, An. subpictus* and *An. fluviatilis* still nothing is determined about the sibling species prevailing in the Thar Desert, and this presents a new pathway for future research.

(4) Sandflies: These are important insects also because of their involvement in a vector-borne infection, viz., tropical leishmaniasis or cutaneous leishmaniasis. Sandflies appear to have been present in the Thar for centuries since they specifically transmit this form of leishmaniasis which is endemic to the Thar Desert, and is transmitted by the species largely endemic to the desert alone, e.g., *Phlebotomus papatasi, P. salehi* and *P. sergenti.* In addition, other species which were encountered there included *Sergentomyia punjabensis, S. babu* and *S. clydei.* Totally eight species belonging to two genera have been recorded from the Thar desert.

Cutaneous leishmaniasis, a self-healing disfiguring infection due to *Leishmania tropica,* has been endemic to the Thar desert due mostly to *P. papatasi* bites, and restricted largely to Sri Ganganagar and Bikaner districts albeit historical records show its prevalence in Jodhpur as well. Canids and rodents (such as desert gerbil, *Meriones hurrianae*) are zoonotic reservoirs. With the advent of the IGNP, cutaneous leishmaniasis, which occurred in both anthroponotic as well as zoonotic types in urban and rural set-ups, in fact showed a clear demarcation of its distribution in the Thar desert due to effects on the habitations as well as distributional and behavioural shift of rodents involved particularly the disease refractory *Tatera indica,* common in agricultural zoine. While the dry type (anthroponotic cutaneous leishmaniasis) is confined to towns and urban agglomeration, the wet type (zoonotic cutaneous leishmaniasis) was prodigiously prevalent in the desert lands recently brought under cultivation and irrigated under the IGNP. If Bikaner could be momentarily considered the central focal point of concentration of the infection before the dawn of canalized irrigation in the Thar desert, it would become quite comprehensible to draw pathways of extension cutaneous leishmaniasis, especially zoonotic type, more prevalent in Bikaner zone, more or lessa within the reach of the IGNP.

(5) Tabanids: These insects deserve special mention in the Thar desert because of abundance of animals, like camels and sheep, besides their involvement in an array of infections, particularly, 'surra', myiasis etc., on one hand, and transmission of infections such as bacterial, viral, helminthic and protozoans (Shri Prakash *et al.,* 2005). Tabanids hitherto collected from the Thar Desert included: *Atylotus agrestis, A. doprsilinea, Tabanus hisarensis, T. striatus, Stomoxys calcitrans* and *Musca crassirostris.* It is understood that since the Thar Desert was in olden types subjected far too often to famines, animal cadavers were a frequent sight, thus making attraction to the tabanids to the putrefying animals.

CONCLUSION

It is more than obvious that insects in the Thar desert, previously only dismally occurring both in number and species during the pre-IGNP era, are now abundantly present, though their site-related prevalence has transformed. Many insects such as termites are threatened to become extinct or displaced due to high moisture in the soil by the IGNP, whereas insects like sandflies bhave dynamically shifted their habitats from the areas currently under IGNP due to erosion of zoonolic habitats of the desert gerbil, *M. hurrianae,* but also the disease (*Cutaneous leishmaniasis*) refractory *T. indica.* But, what mosquitoes, particularly anophelines exhibit in terms of both ecosystem functionality and new habitat adaptations is a unique example of their reproductive plasticity and adaptive prowess in the changed heterogenous desert environments. While the anopheline, *An. stephensi,* inherently adapted to breed in only confined water sources, has adapted to breed in open water bodies as well, another anopheline, *An. culicifacies,* a new entrant in the Thar Desert, and a serious malaria vector, has spread over large areas in the Thar Desert due to extensive, over

10,000 km network of the IGNP, canalization in most 6 desert districts. Mosquitoes such as *C. tritaeniorhynchus*, the vector for Japanese encephalitis, and *Aedes aegypti*, the vector for dengue and chikungunya, have made entries into the xeric rural environments of the Thar. It is significant to note that already both dengue and chikungunya have been reported occurring in the Thar Desert in recent years, and in future the same could happen for Japanese encephalitis. Entry of *Ae. aegypti* into interior rural environments is a saga to be taken note of. The 'tanka', originally meant for storing potable water over long drought periods in the desert, were breeding *An. stephensi* alone. Later on as the conduit water was made available to the desert populations, these tanks were left unattended and uncared which tended to make the long-stored water become turbid due to vegetational growth and other factors. Due to this *An. stephensi* disappeared, but the changed 'tanka' aquatic environment was suitable for *Ae. aegypti* which reigned there completely, thus spreading into the rural Thar desert through millions of such 'tanka' in villages. This simple case of ecosystem transfer and succession of species composition in the newly adapted functional requirements of major insect vectors draw a great deal of significance in disease epidemiology.

References

Christophers, S.R. 1933., The fauna of British India. Diptera III. Family Culicidae, Tribe Anophelini. Taylor and Francis, London, pp. 371.

Covel, G., 1927. A critical review of the data recorded regarding transmission of malaria by different species of Anopheles, with notes on distribution, habits and breeding places. *Indian Medical Research Memoirs* 7: 87-99.

Jaswant Singh, S.A.S., 1933. Some observations on the mosquitoes and sandflies of Rajaputana. Records of Malaria Survey of India 3: 579-581.

Nagpal, B.N. and V.P. Sharma, 1995. *Indian anophelines*. Oxford, New Delhi. 416 pp.

Pruthi, H.S. and D.R. Bhatia, 1952. Pecularities of the insect fauna of the Rajasthan desert and the share of insects in the maintenance of the desert. Proc. Of the Symposium on the Rajaputana Desert, pp. 241-245.

Puri, I.M., 1955. *The distribution of anopheline mosquitoes in India*. Health Bulletin No. 17; Malaria Bureau No. 7, Government of India, 104 pp.

Puri, I.M., 1960. *Synoptic keys for the identification of the full grown larvae of the Indian anopheline mosquitoes*. Health Bulletin No. 16, Malaria Bureau No. 7, Government of India Press, Calcutta, pp. 104.

Ramakrishna, Y.S. and A.S. Rao, 1991. Climatic features of the Indira Gandhi Canal region. (In: Prospects of the Indira Gandhi Canal Project. Ed. I.P. Abrol & J. Venkateshvaralu), ICAR, New Delhi. 1-10 pp.

Rao, A.S., 1996. Climatic changes in the irrigated tracts of Indira Gandhi Canal region of arid Western Rajasthan, India. Annals of Arid Zone 35: 111-116.

Rao, T.R., 1984. The anophelines of India. Malaria Research Centre, New Delhi. pp. 594.

Rathore, N.S. and Madhumita Mandal, 2005. Impact of changing ecology on termite fauna in arid and semiarid regions of Rajasthan. (In:Changing Faunal Ecology in Thar Desert. Eds.: B.K. Tyagi & Q.H. Baqri), Scientific Publishers, Jodhpur, pp. 61-83.

Sikka, D.R. and S.M. Kulshrestha, 2001. Climate and Health studies: Status and scope in India. Centre for Ocean-Land-Atmosphere Studies/Centre for Application of Research on the Environment, Joint COLA/CARE Tech. Rep. No. 5; 66 pp.

Sri Prakash, Vijay Veer, M.J. Mendki, G.P. Gupta and K.C. Verma, 2005. Some new records of mosquitoes (Culicidae) and haematophagous flies of Tabanidae and Muscidae from the Thar Desert (Rajasthan, India). (In: Changing Faunal Ecology in Thar Desert. Eds.: B.K. Tyagi & Q.H. Baqri), Scientific Publishers, Jodhpur, pp. 135-142.

Subbarao, S.K., N. Nanda and K. Raghavendra, 1999. Malariological stratification of India using *Anopheles culicifacies* sibling species complex. ICMR Bulletin 29:75-80.

Tyagi, B.K., 2002. Malaria in the Thar Desert: Facts, figures and future. Agrobios (India), Jodhpur. pp. 165.

Tyagi, B.K. and Chaudhary, R.C., 1997. Outbreak of falciparum malaria in the Thar Desert (India), with particular emphasis on physiographic changes brought about by extensive canalization and their impact on vector density and dissemination. *Journal of Arid Environments* 36: 541.

Tyagi, B.K. & S.P. Yadav, 1996. Malariological and sociological significance of 't 'tanka' and 'beri' in the Thar desert, western Rajasthan (India). Journal of Arid Environments 33:497-501.

Wattal, B.L. and N.L. Kalra, 1962. Regionwise keys to the females of Indian Anopheles. Bulletin of National Society of India for Malaria & other Mosquito-borne Diseases 10: 55-138.

Warren, A., Y.C. Sud and B. Rozanov, 1996. The future of deserts. Journal of Arid Environments 32:75-138.

Chapter - 13

Epilogue

The structure of biological communities and the diversity of life are shaped through specialization in the interaction between living organisms and the environment. Long term detailed studies of patterns of specialization emphasize that extreme host specificity is common in phytophagous insects. One of the most striking aspects of evolving interactions between species is that new associations are being continuously formed within communities, some of them specialized, other evolving over time. Infestation of the same host plant by several species of thrips often results in competition between species and also between different stages of the same species. In order to avoid inter-and intraspecific competition, different species select different sites on the same plant as a measure of resource partitioning. Interspecific competition may have a significant influence on the abundance and distribution of an insect species. As such, the evolution of extreme specialization and reciprocal change within local populations result in different degrees of specialization and adaptation. Specialization has the potential to evolve in different directions in different populations of a species, resulting in the flexibility of specialization. Needless to emphasize that, priorities for conservation must be set on the basis of species richness, irrespective of their impact on ecosystem structure. Key ecosystems should be given priority indicating the need for a diverse range of ecosystems to enable assessment of the role of biodiversity in supporting ecosystem function and resilience. The key process influencing diversity say, of forest patches relate to such biological processes as competition for nutrients, light and water. Depending upon the tree species, understanding vegetation and microbial communities also influence biodiversity. To promote species diversity, the diversity of habitats within agricultural landscapes, through adoption of agroforestry system, maintenance of vegetation in uncultivated corridors appears vital. The development of species rich understories have promoted greater levels of biodiversity and often serve as refuges for threatened species.

Despite all the differences in our approach to biodiversity studies, there is a general agreement that such studies refer to the range of variation in living organisms and their environment, with these levels of hierarchy, viz, genes, species and ecosystem and how ecosystem functioning depends on the level of biodiversity.

To quote Wilson (1992) "The fundamental advance of eighteenth century biology was the descriptions of diversity by organizing species with clusters, according to how closely they resemble one another, eventually developing the heirarchy of classifications. However, by the 20th Century it became more appropriate to describe diversity by the birth of biological organization which are arranged with ecosystem, community, guild, species, organisms and gene leading to two important questions - `how assembled by the certain of biodiversity is ecosystems and alternately how ecosystem functioning depends on the birds of biodiversity". The different chapters in this book provide adequate testimony to this.